T0331010

Supply Chain Risk Management

An Emerging Discipline

Supply Chain Risk Management

An Emerging Discipline

Gregory L. Schlegel
Robert J. Trent

CRC Press
Taylor & Francis Group
Boca Raton   London   New York

CRC Press is an imprint of the
Taylor & Francis Group, an **informa** business

CRC Press
Taylor & Francis Group
6000 Broken Sound Parkway NW, Suite 300
Boca Raton, FL 33487-2742

© 2015 by Taylor & Francis Group, LLC
CRC Press is an imprint of Taylor & Francis Group, an Informa business

No claim to original U.S. Government works

Printed on acid-free paper
Version Date: 20140821

International Standard Book Number-13: 978-1-4822-0597-8 (Hardback)

Visit the Taylor & Francis Web site at
http://www.taylorandfrancis.com

and the CRC Press Web site at
http://www.crcpress.com

Contents

Preface

Perhaps the best way to introduce a book about supply chain risk management (SCRM) is to start with some real although not necessarily uplifting stories. Each of the following occurred in the same week and year during a December holiday season. The names of the companies involved have not been changed to protect the innocent.

Guaranteed On-Time Delivery, Except When It's Not. In its end-of-year edition, *Business Week* magazine prominently featured a cover story about how UPS was going to save Christmas. The magazine chronicled the efforts of the man responsible for making sure all those packages ordered just before Christmas would make their way under the tree in time. Retailers such as Amazon guaranteed that orders placed by December 23 would arrive in time for the big day. This was going to be a defining moment for supply chain managers and online retailers! A convergence of events, however, ensured that Scrooge would have the final say.

What actually happened is a perfect storm that will be studied for many years. While big shippers like Amazon claimed their innocence by announcing that its shipments were given to UPS on time (failures from risk events almost always feature blaming someone else), not enough planes at UPS were available to move such a large number of packages, creating huge bottlenecks.

So, what happened? More consumers than forecast shopped online that holiday season, creating higher-than-anticipated demand. And, only 26 days separated Thanksgiving and Christmas, compared with 32 days the previous year. A great deal of shopping was crammed into fewer shopping days. It did not help that bad weather across much of the United States during this period interrupted package delivery service. Bad weather had a secondary effect of keeping consumers inside where they proceeded to do to even more online shopping. And not surprisingly, many consumers waited until the last minute to place their orders. Why not wait? Retailers such as Amazon guaranteed delivery even though UPS has some fine print stating that delivery is not guaranteed during peak holiday periods. Unfortunately, UPS took a substantial hit to its earnings and reputation.

When Swiping Means Getting Swiped. Target Corp. announced that 40 million customer credit cards were in jeopardy because of a security breach at its point-of-sale store registers. A few days later Target admitted that personal data for up to 70 million customers was also compromised. The retailer told customers they should examine transactions made on their credit and debit cards during a 19-day period and report any fraudulent sales. Making matters worse, credit and debit card accounts stolen during this period reportedly flooded underground black markets, going on sale in batches of one million cards. A fraud analyst at a major bank said his team purchased a portion of the customer accounts from an online store advertised in cybercrime forums. The reporting of this security breach coincided with a subsequent drop in Target's sales, likely due to a loss in customer confidence.

Shortly after the security breach Target, executives announced a set of actions that cost some serious money

Target closed the access point that the criminals used and removed the malware they left behind; hired a team of security experts to investigate the security breach; communicated that its customers would have zero liability for any fraudulent charges arising from the breach; and offered one year of free credit monitoring and identify theft protection to all customers. It's no fun getting swiped.

Heavy Metal Hoarders. A report in *The Wall Street Journal* revealed that banks, hedge funds, commodity merchants, and other investors were hoarding tens of millions of tons of aluminum, copper, nickel, and zinc in a system of hidden warehouses around the world. So what's the big deal? Once hidden in these warehouses, these metals are no longer tracked, making accurate calculations of market supply, something that is needed to determine commodity prices, next to impossible to determine. Producers are bracing for wild swings in metals' prices as speculators withhold data to take advantage of pricing volatility. Market manipulation is likely as metals are controlled by fewer and fewer hands whose interests are likely not aligned with legitimate commodity users.[1]

Toss This Example. In an unfortunate case of how the Internet and social media can place a company's reputation at risk in the blink of an eye, a home security video system captured a FedEx driver tossing a package onto a customer's porch. This might have remained a local event except for the fact that millions of people watched the uploaded video as it went viral. Judging from the driver's throwing technique he is likely the star of his Frisbee golf team.

Welcome to the world of supply chain risk management. It is a world where the end of your day might not be nearly as good as the start of your day. While the examples presented here caused problems at many levels, and we do not want to diminish the harm that came to innocent bystanders, they illustrate that what can happen in a typical week is not always all that typical. As we will discuss, the supply chain world is becoming riskier rather than safer. A survey used to calculate the Allianz Risk Barometer recently concluded for the first time that supply chain risk is now the top concern of global insurance providers. This reinforces our belief that a book about supply chain risk management is relevant and timely. So, how was your week?

SUPPLY CHAIN RISK MANAGEMENT THEMES

As we progress through this book, certain themes are revealed that underlie our view of supply chain risk management. These themes support the basis for everything we present.

- **The financial impact of supply chain disruptions can be devastating but is often not understood until it is too late.** Studies show that, on average, if a publicly held company experiences a moderate or higher risk event, it can expect a 7%–10% reduction in shareholder value. And, approximately 30% of companies that experience a major risk event are out of business within 24 months of the event, and another 25% are out of business after three years.
- **The supply chain management profession has become too comfortable with the deterministic models and tools developed over the last 35 years.** The relatively stable environment of the last 35 years is no longer in existence, and deterministic tools such as forecasting models and sales and operations planning (S&OP) processes have never taken uncertainty into account. Unfortunately, global supply chain growth has resulted in uncertainty, complexity, and risk growing in frequency and severity. The time has come to utilize probabilistic tools that take into account uncertainty in order to manage risk.
- **SCRM is an evolving discipline and will remain so for the foreseeable future.** To be successful in a new global environment, becoming a risk management leader demands mastering four stages of SCRM excellence: visibility, predictability, resiliency, and sustainability. These are part of something we call the 21st Century Supply Chain Risk Maturity Model.

- **Supply chain strategies driven primarily by cost management and delivery improvements are no longer comprehensive enough.** The time has come to make supply chain risk assessments part of the supply chain planning process. Today these risk assessments are still unfortunately more of an afterthought.

- **Showing a hard return on investment for risk management initiatives is a difficult sell.** How do you justify an investment for managing something as vague as a potential risk event? Our view is that traditional financial models are proving to be inadequate when evaluating risk management investments.

- **Social media is the new risk wild card.** A brand built over 50 years can come under attack with a tweet (regardless of whether the tweet is true or not). A negative video on YouTube can go viral in minutes. Social media can amplify the outcome from risk events that may have previously been localized.

- **The risk ledger has two sides**. One side of the risk ledger is the negative side of risk. The other side of the ledger, however, represents opportunity management. It is the upside of risk, as someone's risk is often another's opportunity. Our focus, while recognizing both sides of this ledger, will stress the downside of risk.

- **Supply chain risk is making it to the big leagues.** Companies are placing supply chain risk management verbiage in their 10K and annual reports, something that was rare not too long ago. This illustrates how seriously supply chain risk is being taken at the corporate level. Unfortunately, it also shows how serious the impact can be from supply chain disruptions.

- **Risk heroics must give way to risk prevention wherever possible**. Interviews with leading executives lead us to a clear conclusion. Most companies are tired of responding, sometimes heroically, when a risk event occurs. Increasingly these companies would like to model, anticipate, and even prevent risk events from occurring. The pendulum needs to shift from heroic responsiveness to proactive risk prevention wherever possible. Constantly running around with your hair on fire gets tiring.

- **We need to take a broader rather than narrower view of supply chain risk management.** As a concept, SCRM is similar to Lean and Six Sigma. A narrow view of these concepts considers them mainly as a set of tools and techniques. The broader view, and the one endorsed throughput this book, is that SCRM, like Lean and Six

Sigma, is supply chain–wide, affects an organization's culture, and can have a positive or negative strategic impact.

- **Supply chain risk is increasing, not decreasing.** With globalization expanding at a remarkable rate over the last 20 years, supply chains have moved into areas where they've never operated. Thus, uncertainty, complexity, and risk have grown exponentially. If anyone claims that supply chain risk is decreasing in terms of impact and concern, ask to see their evidence. We will show an abundance of evidence to indicate the contrary.

ORGANIZATION OF THIS BOOK

This book is organized into four sections. The first section sets the stage by positioning our understanding of supply chain risk management. Chapter 1 explains the important concepts and terminology that appear throughout this book. The second chapter provides an overview of the "as is" state of SCRM, an overview that reveals that while most managers appreciate the importance and danger of risk, few organizations are prepared for this new environment. Chapter 3 recognizes that achieving excellence in any area, including risk management, does not happen simply because a company announces its desire for excellence. It also highlights a set of enablers that provide the foundation for effective risk management.

The second section of this book presents a traditional but still important view of SCRM. Here, we address strategic risk (Chapter 4), hazard risk (Chapter 5), financial risk (Chapter 6), and operational risk (Chapter 7). These chapters will describe many approaches for addressing risk within these four categories.

Section III dives into the emerging discipline called supply chain risk management. Chapter 8 addresses fraud, corruption, theft, and counterfeiting; while Chapter 9 presents a set of emerging risk management frameworks. This is followed by two leading-edge topics—using probabilistic models to understand risk (Chapter 10), and using analytics to predict the future (Chapter 11). Chapter 12 presents an emerging set of risk management tools, techniques, and approaches that are broader than what we typically associate currently with risk management. The important topic of risk measurement appears in Chapter 13, and Chapter 14 presents an overview of companies that are well respected in terms of their risk management capabilities. The final section of the book consists of a single chapter that provides a forward-looking perspective in terms of SCRM. This chapter also includes a set of steps for moving a company's risk management agenda forward.

This book also includes an appendix, which presents a risk self-assessment tool that will provide value far beyond the cost of this book. We also provide a web address for free access to this tool.

Although this book is not a novel, we recommend reading the chapters in the sequence they are presented. Rest assured, however, that moving out of sequence will not get anyone in too much trouble.

CONCLUDING THOUGHTS

As we proceed, it is important to keep in mind that risk management capabilities are often relative, which the following narrative illustrates: The CEOs of two competing companies are walking through the woods when they come upon a very large and ornery bear. As the bear roars menacingly, one CEO drops quickly to his knee and begins to tighten his shoelaces. The other CEO says, "What are you doing? You can't outrun that bear!" The first CEO replies, "I don't have to outrun that bear. I only have to outrun you!"

Often in business we only have to run a bit faster than our competitors. The same is true in risk management. While we would always like to anticipate and then prevent risk from happening, when risk events do occur, being faster, flexible, and more responsive than others can make a world of difference. A primary objective of this book is to understand within the domain of supply chain risk management how to run a bit faster and better than the others. Let the journey begin!

ENDNOTE

1. Shumsky, Tatyana. "Heavy Metal Lurks in the Shadows." *The Wall Street Journal*, December 27, 2013: C1.

About the Authors

Greg L. Schlegel, CPIM, CSP, JONAH is the vice president of business development for Shertrack LLC. He has been a supply chain executive for more than 30 years with several Fortune 100 companies and spent seven years as an IBM supply chain executive consultant. Greg was APICS' 1997 International Society President. He is well published and a frequent speaker at conferences, seminars, webinars, and dinner meetings.

Greg has taught operations management at the University of Scranton and has been guest lecturer at Arizona State University, St. Johns University, and Rutgers University. He is presently a member of the Business Analytics Roundtable for Villanova University, a member of the board of advisors for Rutgers University's supply chain undergraduate program, and executive in residence for Lehigh University's Center for Value Chain Research. Greg has taught graduate level supply chain risk management at Lehigh University and has been facilitating supply chain risk management public workshops and the new APICS-supported Supply Chain Risk Certificate workshops around the globe for over three years. He is founder of the Supply Chain Risk Consortium, a group of 13 companies providing education, assessment tools, and consulting services in support of supply chain risk management projects. He teaches enterprise risk management at Villanova in their Executive MBA program. Greg is certified CPIM, CSP in systems, and a Theory of Constraints–certified JONAH. He holds a BS in operations research and computer science from Penn State University and did his graduate work at Lake Forest College.

Greg presently lives in Flemington, New Jersey, with his wife Mariann. He can be reached at schlegel01@earthlink.net.

Robert J. Trent, PhD is the supply chain management program director at Lehigh University. He holds a BS degree in materials logistics management from Michigan State University, an MBA degree from Wayne State University, and a PhD in purchasing/operations management from Michigan State University.

Prior to his return to academia, Bob worked for the Chrysler Corporation. His industrial experience includes assignments in production scheduling, packaging engineering with responsibility for new part packaging setup and the purchase of nonproductive materials, distribution planning, and operations management at a regional parts distribution facility. He has also worked on numerous special industry projects. Bob stays active with industry through research projects, consulting, and training services. He has consulted with or provided training services to 40 government agencies and corporations and worked directly with companies on dozens of research visits.

Bob has authored or co-authored six books and dozens of articles appearing in a range of business publications. He has also co-authored five major research studies published by CAPS Research and has made presentations at numerous conferences and seminars.

Bob and his family reside in Lopatcong Township, New Jersey. He can be reached at rjt2@lehigh.edu.

1

Supply Chain Risk Management Setting the Stage

Floods, earthquakes, tsunamis, tornadoes, and billowing clouds of ash from obscure volcanoes all share something in common. Over the last several years these events have been featured prominently in the news—and each has had the inevitable effect of disrupting the supply chains of entire industries. But these kinds of disruptions were not on the minds of Astellas Pharma executives when thieves stole a trailer from a truck stop containing $10 million of the company's pharmaceutical products. What followed was a lesson in supply chain risk that felt like a swift punch in the gut.

When the accountants had completed their final tabulations, they found that the stolen products represented only a fraction of the losses suffered by Astellas. Based on a recommendation from the U.S. Food and Drug Administration, the company contacted every party in its supply chain, ranging from wholesalers to hospitals, warning them of the stolen drugs. As a preventive measure the company withdrew from the marketplace all drugs with the same lot numbers as those that were stolen. Some of the stolen pharmaceuticals required strict climate control, something the thieves (who were eventually caught) were not too concerned about, making a return of these products a necessity. The loss of this trailer eventually cost the company $47 million, wiping out a large chunk of its North American profit for that quarter.[1]

Welcome to the sometimes unpleasant world of supply chain risk management. This chapter starts our journey into this evolving discipline by setting the stage for important concepts that appear throughout this book. We begin by providing various definitions and perspectives of this thing called *risk*. Next, we present reasons why a focus on supply chain risk management has become a necessity rather than a luxury. This is followed by an explanation of various risk terms and concepts, a categorization of risk, and a presentation of generic risk management approaches.

THE CONCEPT OF RISK AND RISK MANAGEMENT

A logical place to start is to explain what we mean by risk, particularly since this concept can be defined in various ways. One common perspective simply says that risk is a situation involving exposure to danger or loss. Another perspective takes this a step further by adding that risk is the probability or threat of damage, injury, liability, loss, or other negative occurrences that are caused by external or internal vulnerabilities and that may be avoided through preemptive action.[2] Another view states that risk is the effect of uncertainty on objectives. Risk can also be viewed, at least partly, as the inability to capitalize on an opportunity. For our purposes we define risk as the probability of realizing an unintended or unwanted consequence that leads to an undesirable outcome such as loss, injury, harm, or missed opportunity. Warren Buffet once observed that risk comes from not knowing what you are doing.

Most risk observers believe that when a risk becomes a reality, something bad usually happens. Not surprisingly, supply chain managers almost always look at risk in terms of something to be avoided. And to say that most supply chain managers are generally risk averse would be an understatement. Conversely, entrepreneurs look at risk through a different lens. They view risk in terms of upside opportunities and missed opportunities when failing to act. To those individuals, creative risk taking is essential to any goal where the stakes are high. Thoughtless risks are destructive, of course, but perhaps even more wasteful is thoughtless caution, which prompts inaction and promotes failure to seize an opportunity.[3]

Aswath Damodaran, a professor at New York University, writes that every major advance that civilizations have made involves someone willing to take a risk by challenging the status quo. He further states that the most successful firms in any industry actively seek out and exploit risk to their own advantage.[4] He states, "Successful firms, over time, can attribute their successes not to avoiding risk but to seeking out and taking the "right" risks. This perspective views risk as an event or activity that may have an impact on an organization's ability to achieve its objectives or may cause a missed opportunity. The single-minded view that risk is all about avoidance is, in his view, narrow and constraining. It can also be quite paralyzing.

Damodaran's review of risk supports three conclusions that align well with the philosophy of this book. The first is that while some risk definitions focus strictly on the probability of an event occurring, richer perspectives extend this to incorporate a valuation of the consequence of that event. In other words, risk is multidimensional. Throughout this book we will present techniques that consider probability and consequences and model them accordingly. A second conclusion is that in some disciplines a clear distinction is made between a risk and a threat. With this perspective a threat is thought to be a lower probability event while risk is regarded as a higher probability event. Finally, some definitions of risk focus only on the downside of risk, whereas other perspectives are more expansive and consider all variability as risk, including lost opportunities. A company that has more demand for its products than what it is capable of producing appears to have a welcome problem. In reality, the strains placed on that company as it struggles to satisfy demand can affect customer satisfaction, brand reputation, profitability, and even survival.

Each day every company and human being face risk situations. At the individual level, did you drive a car or fly in a plane today? Did you cross a busy street or share the road with cars while riding a bike? Did you eat food at a restaurant where you did not see how the food was prepared? Did you walk down a flight of stairs? Did you step into the shower? Do you have money in the stock market? Did you take an exam without studying? If the answer to even a few of these questions is yes, you have exposed yourself to risk, just like everyone else on the planet. The challenge becomes one of not allowing a fear of risk to paralyze us from pursuing opportunities that are important to our personal and professional advancement. Risk is something we need to manage.

Defining Enterprise Risk Management

It is important to differentiate between enterprise risk management (ERM) and supply chain risk management (SCRM), distinctions that are central to this book. Almost all corporate executives are aware of ERM, a concept that has been around for decades. Using a definition developed by the Aberdeen Group, ERM is the process for effective identification, assessment, and management of all significant risks to an entity. This includes not only the traditional areas of financial and hazard risk, but also larger operational and strategic risks. ERM refers to the people, tools, systems, and structures that are part of a broader framework of Governance, Risk, and Compliance.[5]

Chapter 9 will highlight several ERM frameworks, including the then COSO (Committee of Sponsoring Organization of the Treadway Commission) framework and the ISO (International Organization for Standardization) standards relating to risk as well as Governance, Risk, and Compliance.

Corporate executives have been concerned with enterprise risk for years, particularly at publicly traded companies. The Securities and Exchange Commission (SEC) requires publicly traded companies to identify the material risks they face at the corporate level in Section 1A of their company's 10-K report. Failure to identify these risks can result in claims by shareholders that the company did not adequately warn them of potential risks, which could present some liability to a company.

Risk identification within the 10-K reporting requirements is an important part of the ERM process. Historically, the vast majority of risks identified in the 10-K report related to financial and legal risks. Operating and other supply chain risks simply were not perceived as important enough to be addressed at the ERM or 10-K level. Unfortunately, the world has changed and, from a risk perspective, not for the better.

Table 1.1 identifies the enterprise risks identified in Apple's 10-K report. More than one third of the key risks identified by Apple have supply chain connections or implications (those risks are designated with a check mark); something that is becoming increasingly prevalent as supply chain risks earn the dubious "honor" of making the enterprise risk list. While supply chain managers have been asking for increased attention at the corporate level for years, increasing the number of supply chain–related risks on the 10-K report is probably not what they had in mind. Watch what you wish for.

ERM is traditionally the responsibility of finance, treasury, insurance, and legal groups at the corporate level. In fact, a survey by Accenture revealed that at the corporate level, 98% of organizations have what they consider to be a chief risk officer. And, according to Accenture, 96% of risk management owners report to the CEO.[6] With that said, the chief risk officer is often a dual position. At General Motors, for example, the chief risk

TABLE 1. 1

Apple Enterprise Risk Factors: 10-K Report

- • Global economic conditions could materially adversely affect the company.
- • Global markets for the company's products and services are highly competitive and subject to rapid technological change, and the company may be unable to compete effectively in these markets.
- ✓ To remain competitive and stimulate customer demand, the company must successfully manage frequent product introductions and transitions.
- ✓ The company faces substantial inventory and other asset risk in addition to purchase commitment cancellation risk.
- ✓ Future operating results depend upon the company's ability to obtain components in sufficient quantities.
- ✓ The company depends on component and product manufacturing and logistical services provided by outsourcing partners, many of whom are located outside of the United States.
- ✓ The company relies on third-party intellectual property and digital content, which may not be available to the company on commercially reasonable terms or at all.
- • The company is frequently involved in intellectual property litigation and could be found to have infringed on intellectual property rights.
- • The company's future performance depends in part on support from third-party software developers.
- ✓ The company depends on the performance of distributors, carriers, and other resellers.
- ✓ The company's retail segment has required and will continue to require a substantial investment and commitment of resources and is subject to numerous risks and uncertainties.
- • Investment in new business strategies and acquisitions could disrupt the company's ongoing business and present risks not originally contemplated.
- ✓ The company's products and services may experience quality problems from time to time that can result in decreased sales and operating margin.
- • The company is subject to laws and regulations worldwide, changes to which could increase the company's costs and individually or in the aggregate adversely affect the company's business.
- • The company's success depends largely on the continued service and availability of key personnel.
- ✓ The company's business may be impacted by political events, war, terrorism, public health issues, natural disasters, and other circumstances.
- • The company's business and reputation may be impacted by information technology system failures or network disruptions.
- • The company may be subject to breaches of its information technology systems, which could damage business partner and customer relationships, curtail or otherwise adversely impact access to online stores and services, and could subject the company to significant reputational, financial, legal, and operational consequences.

continued

TABLE 1. 1 (continued)

Apple Enterprise Risk Factors: 10-K Report

- The company's business is subject to a variety of U.S. and international laws, rules, policies, and other obligations regarding data protection.
- The company expects its quarterly revenue and operating results to fluctuate.
- The company's stock price is subject to volatility.
- ✓ The company's business is subject to the risks of international operations.
- The company is exposed to credit risk and fluctuations in the market values of its investment portfolio.
- ✓ The company is exposed to credit risk on its trade accounts receivable, vendor nontrade receivables, and prepayments related to long-term supply agreements, and this risk is heightened during periods when economic conditions worsen.
- The company could be impacted by unfavorable results of legal proceedings.
- The company could be subject to changes in its tax rates, the adoption of new U.S. or international tax legislation, or exposure to additional tax liabilities.

officer is also the company's general auditor. At other companies the chief risk officer may be the chief financial officer (CFO). And at some companies the chief risk officer may be part of the insurance group.

Defining Supply Chain Risk Management

Now that we have a working knowledge of ERM, what is supply chain risk management (SCRM)? The definition partly reflects someone's professional discipline or where they reside in the supply chain. In the information technology space, the National Institute for Standards and Technology defines supply chain risk management as a "multidisciplinary practice with a number of interconnected enterprise processes that, when performed correctly, will help departments and agencies manage the risk of using information technology products and services."[7] MITRE, a private, not-for-profit corporation that provides engineering and technical services to the federal government, defines SCRM as "a discipline that addresses the threats and vulnerabilities of commercially acquired information and communications technologies within and used by government information and weapon systems. Through SCRM, systems engineers can minimize the risk to systems and their components obtained from sources that are not trusted or identifiable as well as those that provide inferior material or parts."[8] A third perspective, and the one that most closely aligns with our philosophy, says that supply chain risk management (SCRM) is "the implementation of strategies to manage everyday and exceptional risks along the supply chain through continuous risk assessment with the objective of reducing vulnerability and ensuring continuity."[9]

One way to view supply chain risk management is to think of it as the intersection of supply chain management and risk management. One thing we know about SCRM is that no standard definition exists. This is one indicator that SCRM is still an evolving discipline. Risk is embedded within so many business disciplines that it should come as no surprise that different groups perceive this concept differently.

WHY FOCUS ON SUPPLY CHAIN RISK MANAGEMENT?

Anecdotal accounts of why supply chain risk management must become a corporate concern are not hard to come by. In fact, we will present dozens of examples that reveal the downside of risk. While natural disasters like hurricanes and floods grab the headlines, the reality is that supply chains face a whole range of risks that most observers believe only to be increasing. A survey by American Productivity and Quality Center (APQC) revealed that 75% of responding companies indicated they were hit by a major supply chain disruption during the two-year period prior to the date of the survey.

A classic example of supply chain risk involves a fire that destroyed an electronics supplier in New Mexico that supplied Nokia and Ericsson with critical components for their phone businesses. The response to this risk event shows the strategic implications of effective (or ineffective) risk management. Nokia's ability to quickly secure components from other sources, compared with Ericsson's lack of preparation for responding to this event, resulted in a dramatic industry shift. Ericsson's supply disruption not only cost the company several hundred million dollars in lost sales, but it essentially ended the company's position as a player in the growing wireless phone business. Chapter 9 will investigate this example in greater detail. Consider some other supply chain risk events:

- A U.S. producer of power tools was surprised to find that the Asian supplier it contracted with to produce its lower-end products began selling those products under its own label in Asia. The U.S. company was further surprised to find that the supplier shared its product designs with other Asian companies. The U.S. producer eventually found itself competing in North America with its own products.

- Some German thieves developed a creative way to steal freight on highways. The thieves position a car in front of a truck to slow it down while another car is positioned next to the truck to prevent it from passing the car in front. Then, a third vehicle pulls up behind the truck and at that point one of the gang members opens the back of the truck to remove cargo. Thieves have used this method to steal cargo more than 50 times.[10]
- Nylon-12 is a critical resin for producing fuel lines and other automotive components. Unfortunately, the resin supply for the entire world is essentially produced in a single facility in Germany. What is even more unfortunate is the explosion that ripped through that plant, taking out half of the world's output in the blink of an eye. Within hours automotive original equipment manufacturers (OEMs) had established crisis management teams to scour the globe for new supply sources.
- Eight heavily armed thieves dressed as police and driving two police vans with flashing lights drove through a hole in the perimeter fence of the Brussels, Belgium, airport and onto a runway. In less than five minutes the thieves opened a plane's cargo door and unloaded 120 packages holding $50 million worth of polished and uncut diamonds. The thieves escaped with the diamonds and are forever embedded in criminal folklore.

We could go on, but you get the idea. Moving beyond anecdotal accounts, an emphasis on supply chain risk management is necessary today because supply chains face many factors that result in higher risk, more so than at any time in modern history. Some of these risk factors are self-inflicted; others are not. IBM researchers have identified a solid set of factors that lead us to a clear conclusion—supply chains are becoming more, rather than less, risky. Table 1.2 summarizes this important set of factors.

Other factors inadvertently expose a company to heightened supply chain risk through unintended consequences. This includes just-in-time delivery and lean systems that result in little to no buffer inventory; a trend toward centralized decision making that may reduce response times and flexibility at local levels; continuous cost reductions that may affect a company's ability to plan and respond to risk events; greater use of single sourcing, which often leaves a company with few supply options and higher supplier switching costs; and widespread outsourcing, potentially leading to a loss of supply chain control. Sometimes we are our own worst enemy.

TABLE 1. 2

Factors That Make Supply Chains Riskier

- Increased globalization through outsourcing, which stretches end-to-end supply chains
- Additional regulatory compliance imposed by government entities, further complicating international trade (such as C-TPAT and SEC conflict mineral reporting requirements)
- Increased levels of economic uncertainty and market volatility, which create additional variability in demand and supply and make it more difficult to accomplish demand–supply planning
- Shorter product life cycles and rapid rates of technology change, which increase the risk of inventory obsolescence
- Demanding customers that create additional time-to-market pressures by requiring better on-time delivery, higher order fill rates, and improved service level efficiencies
- Supply side capacity constraints, making it more difficult to meet demand requirements
- Natural disasters and external environmental events, which affect global supply chains
- Complex networks of suppliers and third-party service providers, as well as large interdependencies among multiple firms, which increase the need to coordinate risk

---

A study by the Aberdeen Group identified some good reasons why a company should make SCRM an embedded part of its corporate culture. First, a need to protect an organization's brand and competitive advantage is a strategic necessity. Risk events have a nasty way of affecting brand value quickly. Simply think about how stories, whether they are true or not, can impact the value of a brand. Next, the increasing volatility of the global economic environment and markets is resulting in greater risk exposure. Third, corporate mandates to institute and/or improve risk management and governance programs are only going to increase. And, a growing need to comply with new or changing regulatory requirements is forcing a greater emphasis on risk management. Finally, constant pressure to improve shareholder and customer confidence while trying to reduce costs may result in actions that result in greater risk exposure, such as searching for suppliers in untested emerging supply markets.

A range of surveys and studies conclude that supply chain risk is growing. To disregard what has become obvious is short-sighted and dangerous. We can easily cite source after source that concludes essentially the same thing—supply chain risk and its impact on corporate performance continues to grow. It would be challenging to argue that supply chains are, on average, becoming less risky.

Some SCRM Observations

Extensive experience and research enables us to make some observations about the state of risk management. (Chapter 2 will provide a more in-depth presentation of the "as is" state). Perhaps most importantly, most observers have concluded that the potential impact of risk has increased over the last 15 or 20 years. In one survey, almost 75% of risk managers say that supply chain risk levels are higher than in 2005. More than 70% say the financial impact of supply chain disruptions has also increased.[11] And, there is no question that supply markets have become more volatile. The size of fluctuations in commodity prices has more than tripled since 2005 compared with the period of 1980–2005, based on International Monetary Fund data. If you really think about this hard enough, you might just get depressed.

We can also conclude that too many firms are not prepared to handle the supply chain risks that may come their way, even though most managers understand that supply chain risk is a growing concern. While ERM has been at the forefront for many companies, SCRM has been more of an afterthought. A recent study revealed that for firms with less than $500 million in annual revenue (which is the vast majority of companies), only 25% take a proactive approach to risk management.[12]

Another observation is that while many risk categorizations and topologies exist, a convergence appears to be happening around the key categories of supply chain risk—a convergence this book uses. Finally, as it relates to mitigating or lessening the impact of risk events, we tend to see the same set of standard approaches that fail to reflect bold or innovative thinking. While "blocking and tackling" will always be important, it is time to see a bit more creativity and sophistication within the SCRM arena. Later chapters will look at some more advanced SCRM approaches.

Why Aren't We Prepared for SCRM?

The reasons why so many firms are not prepared to manage supply chain risk effectively are varied. We cannot ignore what is perhaps the most likely reason of all—risk management has simply not been a part of the supply chain domain. Why would we focus on something that is not considered all that relevant? It is easy to view the efforts put forth toward risk planning as a big exercise in busy work. This may not be the kind of work that gains personal recognition and promotions.

A study by the Supply Chain Council (SCC) identified a set of barriers that affect the practice of supply chain risk management. One barrier is the tendency of senior management to focus on risk management only during times of crisis, something that needs to shift from responsiveness to prevention. A second barrier is that SCRM requires many functions to cooperate, something that is challenging even on a good day. Third, the study concluded that SCRM responsibilities are typically added to existing staff responsibilities. While everyone should be a risk management stakeholder, adding responsibilities to existing duties clearly creates a competition for resources, a competition that SCRM will often lose. Next, the increasing complexity of products, divisions, regions, and supply chains makes a coordinated SCRM effort more of a challenge. A final barrier is that a partial effort to SCRM dilutes the perceived need for a real and sustained risk management effort. A "close enough is good enough" attitude toward SCRM often prevails. These barriers will clearly affect the state of SCRM.

SOME IMPORTANT RISK CONCEPTS

A working knowledge of some important risk concepts is essential when talking about SCRM, particularly since these concepts are mentioned repeatedly throughout this book. We also do not want someone to appear ill-informed when talking about risk management with others. Part of understanding risk management is having a working knowledge of the terms and concepts that populate this body of knowledge. The following presents some important terms and concepts that will help you speak the language of a risk manager.

Risk Event

An important distinction exists between *risk* and *risk events*. Every day we face hundreds of risks with various probabilities attached to them (although we rarely quantify those probabilities). But, and this is important, a risk is relatively harmless until it happens. There is always a risk that someone will fall off a roof when they are working on their house. Until that person actually takes the plunge, the risk of falling remains simply a risk. If the person falls, the risk is now a *risk event*. A risk event is simple to conceptualize—it is a risk that has become a reality. Formally defined, a risk event is a discrete, specific occurrence that negatively affects a decision, plan, firm, or organism.[13]

Risk events are not only episodic, temporary occurrences. Risk events can be continuous, particularly if they relate to operational performance problems. Any supply chain performance problem that is ongoing presents continuous risk to multiple parties in a supply chain.

A word of caution is in order here. A tendency exists to identify a grab bag of risk events and then label each event as a risk category. This is generally an unorganized way to approach risk management. Late supplier deliveries or supplier quality problems might comprise two such categories even though they are risk events. Risk events should be organized and placed into broader risk categories. In the supply chain space a number of risks might relate to financial risks, for example, and therefore should be placed under a financial risk category. Subcategories of financial risk may then be developed that include supplier financial risk, working capital risk, or currency risk. A later section will present risk typologies.

Risk Exposure and Vulnerability

Risk exposure involves the quantified potential for loss that might occur as a result of a risk event. The risk exposure value is often the outcome of a comprehensive risk analysis that uses algorithms to combine risks according to their probability of occurring against the potential loss if the risk occurs. A company that can seamlessly switch production between multiple supplier locations has less risk exposure to a supply disruption compared with a buyer that has access to only a single production location. Even before a garment factory collapsed in Bangladesh, killing 430 workers in the country's worst apparel-industry accident, major buyers such as Walmart and Levi Strauss had ceased doing work with vendors who operated in multistory buildings. The risk exposure from these operations was simply too great.[14]

For our purposes we view risk exposure and vulnerability as closely related concepts, although vulnerability tends to be a less quantified concept. We are vulnerable to something if we are susceptible to harm or injury. Anyone who has built a house on an earthquake fault will grasp the concept of vulnerability to earthquakes. Or, someone traveling to certain parts of the world without getting proper vaccinations should appreciate being more vulnerable to diseases. In the information technology (IT) world, vulnerability refers to the security flaws that allow a successful system attack by hackers.[15] IT vulnerability is an important concept because supply chains today are increasingly information enabled.

Risk Resilience

Risk resilience is becoming one of the most researched and discussed topics in supply chain risk management. At a basic level, *resilience* refers to the ability to recover from or adjust to misfortune or change.[16] It represents the ability of a company and supply chain to "bounce back" after an event. While the concept of resilience has been studied scientifically in development psychology and ecosystems for many years, it is still an emerging topic in SCRM. Even in well-developed disciplines the definitions of resilience are often contradictory and confusing.[17]

A good example of pursuing resiliency as an objective comes from the utility industry. As utilities work to storm-harden their networks (a form of risk prevention), some are also investing in technology to recover faster from outages (risk responsiveness or mitigation) through an approach called the "smart grid." New systems use advanced technology to pinpoint problems, reroute power around problem areas, and identify where repair crews need to go first to get the most customers restored the fastest.[18] One emerging technology cuts off power at the spot where a tree falls into a power line and then reroutes electricity so nearby customers still retain power. Using a boxing metaphor, resiliency means being able to take a punch and still be standing.

A second resiliency example involves offshore oil exploration in the Gulf of Mexico. It became obvious that following the 2010 explosion at BP's Macondo well an array of new and complex regulations would emerge addressing offshore drilling safety. And that is exactly what happened. Some observers predicted that drilling in the Gulf of Mexico would not recover for years, if ever. But that does not seem to be the case. In the words of one analyst, "Bottom-line, Gulf of Mexico oil production is in considerably better shape than even the most ardent optimists envisioned following Macondo."[19] Part of the reason for such optimism is the oil industry's resiliency as it learns to live with stricter safety oversight and slower permit reviews. Estimates indicate that by 2022 oil output from the Gulf of Mexico will be 28% higher compared with current levels.

Risk Appetite

Risk appetite reflects the degree of risk that an organization or individual is willing to accept or take in pursuit of its objectives. This can be measured in terms of both quantitative and qualitative dimensions. Some also refer to this concept as *risk tolerance* or *risk propensity*, a topic that is well grounded in the financial community.

Finance experts view risk appetite as reflecting the type of risk that an institution or individual is willing to undertake in pursuit of a desired financial performance. Clearly, someone who invests in derivatives rather than guaranteed government bonds (assuming they are not Greek bonds) has a higher appetite for risk. When an organization or individual has a low risk appetite, we say they are *risk averse*. As it pertains to supply chain risk, we can safely conclude that most organizations tend to be risk averse. Remember, the typical supply chain professional looks at risk in terms of loss or harm.

Complex models have been developed to identify risk utility functions. Utility functions transform monetary values (payoffs and costs) into utility values that specify preferences for various monetary payoffs and costs. This encodes a company or individual's attitude toward risk. A time-consuming step when developing utility functions is to assess a company's or individual's attitude toward risk. At the company level, this assessment is part of a dialogue between the board of directors and senior management and includes factors such as business model aspirations, institutional principles, shareholder expectations, and core competencies.[20]

An analysis by *The Wall Street Journal* concluded that the United States is becoming more risk averse (i.e., a lower risk appetite) as a nation compared with previous periods. If this is true it does not bode well for the longer-term growth prospects of the U.S. economy as fewer individuals start new ventures. *The Wall Street Journal* analysis concluded that three shifts are causing Americans to become more risk averse, an aversion that will result in fewer new businesses being created and a reluctance to change jobs or move to take advantage of new opportunities. These shifts include an aging population (older citizens are not known as risk takers), the emerging dominance of large corporations in many industries that shuts out new players and ideas, and a reluctance of venture capitalists to invest in new opportunities. As one observer says, "The pessimistic view is we've lost our mojo."[21] At a national level we need risk takers to grow the economy through innovation and change.

Risk Analysis or Assessment

Risk analysis, also called risk assessment, is the process of qualitatively and quantitatively assessing potential risks within a supply chain. At a basic level risk analysis involves identifying risks and then evaluating or mapping these events, at a minimum, across two dimensions. These dimensions include the probability of a risk occurring and the impact if the risk were to become a risk event. Some techniques will score the two dimensions and multiply them together to arrive at an overall risk score. Chapter 13 will discuss some validity issues related to this approach.

In the financial sector, risk analysis refers to the uncertainty of forecasted future cash flows streams, variance of portfolio and stock returns, statistical analysis to determine the probability of a project's success or failure, and possible future economic states. Remember, risk analysis (and risk management) is far more evolved in the financial community compared with the supply chain community.

Risk Response Plan

A risk response plan is a logical extension of a risk analysis. The risk plan is a document that defines known risks and includes descriptions, causes, probabilities or likelihood of risk occurrence, costs, and proposed risk management responses. A word of caution is in order here. We have all been presented with (or assigned to write) the dreaded 125-page report that no one will ever read. In the old days this report would sit on a shelf in someone's office collecting dust. Now, these reports sit in electronic directories collecting virtual dust. A risk response plan should be a crisp, actionable document that is not someone's idea of busy work.

Risk Compliance

Risk compliance includes the internal activities taken to meet required or mandated rules and regulations, whether they are governmental, industry specific, or internally imposed. Companies have always had compliance requirements relating to financial reporting, environmental compliance, and a host of other areas. At an organizational level, compliance is achieved through management processes that (1) identify applicable laws, regulations, contracts, strategies, and policies; (2) assess the current state of compliance; (3) assess the risks and potential costs of noncompliance against the projected expenses to achieve compliance; and (4) prioritize, fund, and initiate any corrective actions deemed necessary.[22] While compliance reporting requirements have been around for many years, the hazard events of the last 15 years have brought about new compliance requirements, particularly in the area of international supply chains.

Risk Governance

Risk governance includes the frameworks, tools, policies, procedures, controls, and decision-making hierarchy employed to manage a business from a risk management perspective. At times the governance structure includes a chief risk officer, who is normally identified as the person responsible to coordinate and oversee the risk management process and approve reports to the corporate audit committee of the board of directors. Chapter 3 will address the pros and cons of designating chief risk officer.

The risk concepts presented here are certainly not the only ones that comprise the vocabulary of SCRM. They are, however, the more important ones. It would be difficult to proceed with our risk discussion without having this working knowledge of risk terminology.

CATEGORIZING RISK

While various frameworks categorize the domain of supply chain risk, no standard agreement exists regarding what these categories should be. Any categorization scheme should identify broader risk categories and then place specific risks within those categories. One perspective classifies supply chain risk into nine categories—design; quality; cost; availability; manufacturability; supply; financial; legal; and environmental, health and safety.[23] We think that a more simplified approach might better suit our needs.

Perhaps the most logical way to look at supply chain risk is to consider the four categories that define enterprise risk management—strategic, hazard, financial, and operational risks. While some frameworks present more categories, the thriftiness of these four categories is a virtue. The following describes these categories.

Strategic Risk. For something to be strategic, it must be necessary to or important in the initiation, conduct, or completion of a strategy or strategic plan. Strategic risks are those risks that are most consequential to an organization's ability to carry out its business strategy, achieve its corporate objectives, and protect asset and brand value. Chapter 4 explores strategic risk in detail.

Hazard Risk. This category of risk pertains to random disruptions, some of which involve acts of God. This category includes bellowing ash from a volcano in Iceland, a tsunami that devastated Japan, serious floods in Thailand, and a super storm named Sandy that affected the eastern United States. This category also includes fires and malicious behavior such as accidents, product tampering, theft, and acts of terrorism. Hazard risk is normally what we think of when we purchase insurance as a form of risk protection. Chapter 5 addresses this risk category.

Financial Risk. Financial risks relates to the internal and external financial difficulties of the participants within an integrated supply chain. While we can make the argument that all supply chain risk events eventually have financial risk implications, we categorize a risk as financial when the primary and immediate effect of the risk, rather than a subsequent or secondary effect, is financially related. Chapter 6 explores financial risk in detail.

Operational Risk. Operational risk arises from daily operations. By far a disproportionate set of supply chain risks will be categorized as operational since this category includes internal and external quality problems, late deliveries anywhere in the supply chain, service failures due to poorly managed inventory, problems related to poor forecasting, and a thousand other events related to operational performance failures. Chapter 7 addresses operational risk specifically.

Other Ways to Look at Risk

A somewhat different way to look at risk is according to a three-category system that categorizes risks as systemic, event, or idiosyncratic.[24] Systemic risks pertain to widespread risks that impact most players in an industry. Chinese wage inflation and currency reevaluations are risks that will affect a large number of players from many different industries. Event risks include narrow or localized events that impact participants selectively. An earthquake in Taiwan, for example, may selectively impact semiconductor foundry operations. Or, a tornado in Oklahoma only impacts directly a certain part of the United States. Idiosyncratic risk pertains to highly localized events that impact very few players. A delayed truck delivering goods to a single retail store is an example of a risk that has a limited affect in terms of its impact.

Still another way to look at risk involves hard versus soft risks. Hard risks are easily measurable and tangible, such as risks that affect assets, inventory, and facilities. With hard risks a company can identify reasonably precise losses if a risk materializes and a reasonable history of occurrences and probability exists. Soft risks are more difficult to measure or identify.

Because soft risks are usually present to some degree, they increase the overall probability of risk occurrence but in ill-defined or imprecise ways. An analogy here involves total cost models. Some costs are easily identifiable and quantifiable (transportation costs and unit price, for example) while other costs are "hidden" and difficult to calculate (the cost of communication and time-related problems when dealing with remote Chinese suppliers). These hidden costs (which are analogous to soft risks) still increase the true total cost, although in ill-defined or imprecise ways.

Still an additional way to think about risk is in terms of known and unknown risks. Known risks are specific risks that we have encountered previously or can foresee or anticipate with a reasonably good estimate of occurrence. During risk analysis and planning known risks are good candidates for practicing risk prevention. Unknown risks consist of unforeseen combinations of outcomes or events that produce a risk. This includes unexpected or unanticipated surprises. Managing unknown risks will benefit from strong risk mitigation plans.

GENERIC RISK MANAGEMENT APPROACHES

Literally hundreds of activities, tools, and approaches have the potential to be part of a company's risk management portfolio. At a very high level we can organize these approaches by their primary risk objective, which includes mitigating, avoiding, preventing, accepting, or sharing risk.

Risk Mitigation

Some will use the term *risk mitigation* to describe almost everything that is undertaken in the name of risk management, including preventive actions. According to its most basic definition, *mitigate* means to lessen the impact of something. That "something" could be the effect of a risk event such as a supplier fire or supply chain quality problem. In a broader sense, mitigation can also be the result of action taken to either reduce the likelihood of a risk occurring or minimize the extent of its impact. This broader perspective is why many sources refer to just about any risk management initiative as risk mitigation.

When speaking in broad terms about how to manage risk, we will use the term *risk management* rather than *risk mitigation*. We view mitigation more in terms of responding to risk events rather than preventing risk. Prevention is a risk management response that is largely separate from mitigation. Mitigation or risk responsiveness is essential when unknown risks are present. Not everything can be anticipated or prevented.

Risk Avoidance

Avoidance involves exiting those activities that give rise to a risk. A company may decide (and many have) that sourcing a material from a certain supplier is too risky, so it avoids that supplier. Or, a certain line of products is not earning enough profit, so a company decides to stop making those items (or sell the brand to another company). With avoidance, a company has made a conscious decision to reduce, perhaps even eliminate, its risk exposure.

Risk Prevention

Prevention involves taking action to ensure that a risk does not become a risk event or, if it does become an event, that it will have an inconsequential effect. This approach to managing risk is often preferable when dealing with known risks. Prevention is different from avoidance in that a company did not exit something as a means of addressing risk. We expect a greater focus on prevention as supply chain managers become more focused on anticipating and averting risk rather than experiencing and responding to risk. We have heard numerous business leaders comment that their firms, while good at responding to risk events, need to do a better job preventing the risk event in the first place.

At a personal level most of us understand the concept of prevention, particularly as it relates to our personal health. We know that if we watch our weight, exercise, avoid smoking and excessive alcohol, avoid harmful drugs, drive prudently, and eat correctly, we may prevent a host of serious ailments.

We will clearly differentiate between mitigation and prevention activities as we try to anticipate and then take steps to ensure that a risk does not become a reality. A primary argument put forth in this book is that most companies are conversant or even qualified in terms of risk responsiveness (mitigation) but not nearly as advanced when it comes to prevention.

Risk Acceptance

Acceptance means to take on and assume a risk. SCRM may not be a priority at a company, so therefore no specific risk management action is taken. In this case acceptance occurs essentially by default. A second reason for risk acceptance is that a cost/benefit analysis reveals that the cost of addressing a risk outweighs the expected impact of the risk. A third reason is that no practical way exists to prevent, share, or mitigate the risk. This is usually an acknowledgment that, at least in the short run, no viable action or alternative is available that will effectively address a risk. No practical choice exists except to assume the risk.

Risk Sharing

Risk sharing involves transferring or sharing a portion of a risk to reduce or mitigate it. Sharing product development costs with suppliers or buying insurance is a risk-sharing method. We all practice risk sharing when we buy home, car, or life insurance. Financial managers enjoy the benefits of risk sharing when they engage with other traders to hedge commodities and currencies. And supply chain managers often write contracts with suppliers that feature some level of currency or commodity risk sharing.

Risk pooling is an important form of risk sharing. A risk pool involves insurance companies controlling the risk of insuring against catastrophic events or extending insurance to individuals or businesses that are likely to create sizable claims. If a claim arises from a natural disaster or catastrophic weather event, the companies spread losses among all members of the risk pool. Single members of the risk pool are protected from large claims that would bankrupt the insurance company, leaving their claimants with nothing.[25]

Prevention versus Responsiveness

Often times we look at risk simply in terms of trying to prevent an event from occurring versus responding after the fact. While prevention sounds like an ideal way to approach risk management (who would not want to prevent problems?), not all risks can be predicted or anticipated, particularly hazard risks. What we see today at best practice firms is a combination of preventive actions wherever possible combined with responsiveness plans that support risk resiliency. The bottom line in the prevention-versus-responsiveness debate is that an increase in known risks favors pre-risk preventive approaches. And an increase in unknown risks favors post-risk response agility. How a specific company approaches risk management will be affected by this mix of known and unknown risks.

CONCLUDING THOUGHTS

This chapter starts our journey through the evolving world of supply chain risk management. Something to remember as we progress through later chapters is that companies that are effective at managing risk are supported by a corporate culture and a foundation that promote a company-wide approach to risk management. A culture that stresses risk awareness and management must then be supported by a set of enablers and sophisticated tools and techniques, and perhaps most importantly, the ability to quantify the value of risk management efforts. When the right corporate culture is supported by effective tools, techniques, measures, and skills, a company can engage in thoughtful risk taking rather than being paralyzed by an irrational fear of risk.

Summary of Key Points

- One common perspective for defining risk simply says that risk is a situation involving exposure to danger or loss. A broader perspective takes this a step further by adding that risk is the probability or threat of damage, injury, liability, loss, or other negative occurrences that are caused by external or internal vulnerabilities and that may be avoided through preemptive action.

- It is important to clarify the distinction between enterprise risk management and supply chain risk management. Almost all corporate executives are aware (or should be aware) of ERM, a concept that has been around for decades. SCRM is an evolving discipline.

- SCRM is the implementation of strategies to manage everyday and exceptional risks along the supply chain through continuous risk assessment with the objective of reducing vulnerability and ensuring continuity. One way to view SCRM is to think of it as the intersection of supply chain management and risk management.

- An emphasis on SCRM is necessary today because supply chains face many factors that result in higher risk, more so than at any time in modern history. Some of these risk factors are self-inflicted; some are not.

- A working knowledge of some important risk concepts is essential when talking about SCRM, including what is a risk event, risk exposure and vulnerability, risk governance, risk resilience, risk appetite, risk compliance, risk response plan, and risk assessment and analysis.

- While various frameworks are available to categorize the domain of supply chain risk, no standard agreement exists regarding what these categories should be. A logical way to look at risk is to consider the four categories that have defined enterprise risk management for many years—strategic, hazard, financial, and operational risks.

- Literally hundreds of activities and approaches have the potential to be part of a company's risk management portfolio. At a very high level we can organize these approaches by their primary risk objective(s), which includes mitigating, avoiding, preventing, accepting, or sharing risk.

- When the right corporate culture is supported by effective tools, techniques, and measures, a company can engage in thoughtful risk taking rather than becoming paralyzed by an irrational fear of supply chain risk.

ENDNOTES

1. Grushkin, Daniel. "Cargo Theft: The New Highway Robbery." *Bloomberg Business Week*, May 26, 2011. Accessed from www. businessweek.com/magazine/content11_23/b4231072707549.htm.
2. Accessed from http://www.businessdictionary.com/definition/risk.html.
3. Blair, Gary Ryan. as quoted in www.brainyquote.com.

4. Accessed from http://people.stern.nyu.edu/adamodar/pdfiles/valrisk/ch1.pdf.
5. Hatch, David and Cindy Jutras, "The Executive Enterprise Risk (ERm) Agenda," Aberdeen Group, September 2010. p. 5.
6. Teach, Edward. "The Upside of ERM." *CFO*, November 2013: 44.
7. Accessed from http://www.nist.gov/itl/csd/supply-112712.cfm.
8. Accessed from http://www.mitre.org/work/systems_engineering/guide/enterprise_ engineering/se_for_mission_assurance/supply_chain_risk_mgt.html.
9. Wieland, A., and C. M. Wallenburg. "Dealing with Supply Chain Risks: Linking Risk Management Practices and Strategies to Performance." *International Journal of Physical Distribution & Logistics Management*, 42, 10, (2012): 887–905.
10. Accessed from http://securingyoursupplychain.com/theft-of-cargo-from-moving-trucks-reported-in-germany/.
11. Favre, Donovan, and John McCreery, "Coming to Grips with Supplier Risk." *Supply Chain Management Review,* 12, 6 (September 2008): 26. Citing statistics from Marsh, Inc. and *Risk & Insurance* magazine.
12. *Industry Week* risk study. May 2011.
13. Accessed from http://www.businessdictionary.com/definition/risk-event.html.
14. Otto, Ben, Joanna Sugden, and Christina Passariello. "Before Dhaka Collapse Some Firms Fled Risk." *The Wall Street Journal*, May 3, 2013: A7.
15. Accessed from http://www.techrepublic.com/blog/it-security/understanding-risk-threat-and-vulnerability/.
16. Accessed from www.merriam-webster.com.
17. Ponomarov, Serhiy, and Mary Holcomb. "Understanding the Concept of Supply Chain Resilience." *The International Journal of Logistics Management*, 20, 1 (2009): 124–125.
18. Smith, Rebecca. "Getting 'Smart' on Outages." *The Wall Street Journal,* November 8, 2012: B6.
19. Fowler, Tom, "After Spill, Gulf Oil Drilling Rebounds." *The Wall Street Journal*, September 21, 2012: B1.
20. Anonymous. "Institutions Need to Better Understand their Risk Appetite." *The RMA Journal*, 92, 6 (March 2012): 38042.
21. Casselman, Ben. "Risk-Averse Culture Infects U.S. Workers, Entrepreneurs." *The Wall Street Journal*, June 3, 2013: A1.
22. Accessed from www.wikipedia.com.
23. Zsidisin, George, Alex Panelli, and Rebecca Upton, "Purchasing Organization Involvement in Risk Assessments, Contingency Plans, and Risk Management: An Exploratory Study." *Supply Chain Management: An International Journal*, 5, 4 (2000): 189.
24. Wallingford, Jeff, and Ron Keith. "Supply Chain Risks in a Newly Flat World." *Supply Chain Management Review*, 16, 6 (November 2012): 6.
25. Accessed from http://www.ehow.com/about_6521384_risk-pooling-insurance_. html?ref=Track2&utm_source=ask.

2

Supply Chain Risk Management
The As-Is Landscape

In this chapter we provide an overview of the growth of supply chain risk management (SCRM) during the last few years and provide additional insight into who is talking about SCRM, what they are saying, and the tone of the dialogue. We will provide a summarized view of SCRM topics, announcements, articles, reports, and survey results that help define the state of SCRM today. We will also share a perspective on SCRM through a different prism: the Four Pillars of SCRM. We will then provide a maturity level for each pillar. The chapter concludes with discussion of SCRM adoption.

A CHRONOLOGY OF SUPPLY CHAIN RISK MANAGEMENT

Let's start by providing a sense of the growth of SCRM in terms of awareness, discussions, articles, papers, surveys, solutions, tools, and more. During our research for this book we compiled hundreds of reference works in order to provide a comprehensive, end-to-end perspective of SCRM. Using these reference works and our experience, we have compiled a chronology of the growth we've witnessed and the tone and tenor of the dialogue.

While many risk management studies and surveys were conducted during 2000–2010, our focus starts with the latter part of that decade. Most supply chain executives became interested in SCRM after the financial meltdown of 2008. Company after company watched, almost helplessly, as customer orders were canceled, suppliers entered bankruptcy, and commodity markets became increasingly volatile. For many, 2008 was the genesis of their risk management efforts.

2009

Most operations management and supply chain management professionals are familiar with the International Organization for Standardization (ISO) Group and its set of standards that revolve around quality control and process reliability, such as ISO 9000. In 2009, the ISO Group delivered its first set of standards directly relating to supply chain risk, which was a major recognition of the importance of SCRM. These standards include ISO 73 and ISO 31000. ISO 73 profiles the vocabulary and taxonomy of risk within the supply chain, and ISO 31000 provides insight into the principles, practices, and guidelines to effectively identify, assess, mitigate, and manage supply chain risk.

A key report during this period was Accenture's 2009 *Managing Risk for High Performance in Extraordinary Times.*[1] This global report, which surveyed 260 CFOs, chief risk officers, and other executives across 21 countries, came right after the 2008 financial meltdown. This report explored the details of what was going on around the globe relative to risk and its impact on businesses and supply chains. It was one of the first comprehensive analyses of the cause and effects of risk to global supply chains. This report revealed that 85% of executive respondents said their company needed to overhaul its approach to risk management; 40% said their companies already had increased or would increase their investments in broader risk management capabilities; 41% stated their risk management costs increased by at least 25% over a three-year period; and only 27% said their risk management function was involved, to any great extent, in objective setting and performance management.

Accenture concluded that steps needed to be taken to manage supply chain risk to protect a company's competitive advantage, reputation and branding, and credit ratings; to maintain positive comments by analysts; and to ensure access to capital over time at a reduced cost. Overall, Accenture reported that many companies had a long way to go before they should feel comfortable about their company's state of risk management.

AMR released a report on risk in U.S. manufacturing to get a sense of how businesses feel about the future and what they saw in terms of risk. The top three concerns of U.S. manufacturers were supplier quality failures, commodity price volatility, and intellectual property infringement. In particular, supplier solvency was a key concern. Furthermore, respondents said that doing business with Chinese suppliers was contributing the most to overall risk, scoring the highest in 11 out of 15 risk categories evaluated.

2010

In 2010, we began a course for a master's of business administration degree by explicitly profiling the emerging elements of supply chain risk management. The initial courseware was made up of articles, survey outcomes, reports from the insurance industry, and the basics of good risk management. As the course grew in maturity, a Supply Chain Risk Assessment Tool emerged that codified the effects of supply chain maturity as it relates to mitigating risk. The basic premise was that as the supply chain matures, the inherent risks faced by that supply chain diminish.

The Supply Chain Risk Assessment Tool, featured in Figure 2.1, encompasses about 100 questions-of-discovery about a company's supply chain across 10 tenets of the entire supply chain. It is a perception-based tool that captures the complexion of a company's operating environment relative to a maturity model and then develops an inherent risk factor for every tenet. Risk factors are then plotted inside a spider diagram for visual presentation using a red, yellow, and green designation. There will be more discussion about this tool later in the book.

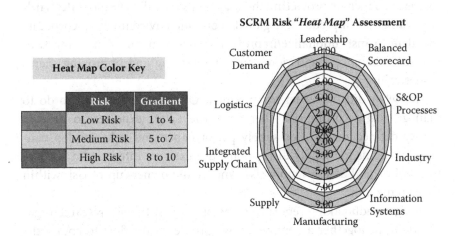

FIGURE 2.1
Tools: supply chain risk assessment.

During 2010, the Massachusetts Institute of Technology (MIT) Center for Transportation and Logistics produced its preliminary findings from the Global Supply Chain Risk Management research project. The survey collected 1,400 usable responses from companies in 70 countries. The project's primary goal was to understand if regional and cultural differences affect how people think about and manage supply chain risks. This survey was a seminal point in the birth of SCRM. The final report covers differences in attitudes about risk, differences in risk frequencies (internal and external events) and priorities, and differences in practices across countries and industries.[2] It is one of the few reports that approached risk from a cross-cultural perspective.

The MIT report concluded that respondents showed a marked preference for risk prevention as opposed to planning and response, but that was influenced by national culture. Furthermore, most respondents indicated that risk planning and prevention should be carried out centrally in an organization, while actual responses to most risk events should be managed locally. Respondents were also asked to rank the top supply chain disruptions for which their company should prepare. A breakdown in supply quality, supplier financial failures, and internal process failures topped the list.

Aberdeen Group also produced a comprehensive report on SCRM. Their findings further supported the notion that SCRM is evolving as a concept.[3] One of the key findings from this study was an analysis of the reasons behind pursuing an aggressive risk management agenda. The top reasons include protecting the organization and its brand, safeguarding against an unpredictable global economic environment, a corporate mandate to institutionalize/improve risk management, complying with new and changing regulations, and maintaining and improving shareholder value.

And what did the Chief Procurement Officers (CPOs) plan to do to manage risk according to Aberdeen? The CPOs expected to develop clearly defined metrics for supply performance and risk, develop contingency plans for supply disruptions, use external information services to monitor and assess supply risk, and define ownership of risk within the organization.

This groundbreaking report was one of the first times a research organization put together a comprehensive profile in an effort to codify the elements of a good SCRM journey. Figure 2.2 illustrates the actions items, capabilities, tools, and techniques that will help ensure success within the

Pressures	Actions	Capabilities	Enablers
– Protect the organization and its brand/ competitive advantage – Unpredictable global economic environment	– Implement processes aligning risk management with compliance/governance – Develop supply risk mitigation strategies – Build risk aware culture throughout the organization	– Standard risk management policies and procedures across the enterprise – Management accountability for risk management activities within overall governance framework – Transparency in communication of risk information (i.e., policies including escalation criteria, procedures, practices & thresholds) – Risk information integrated into core decision-making (i.e., strategic planning capital allocation and performance management)	– Enterprise Risk Management (ERM) platform/software – Enterprise governance, risk, and compliance (GRC) platform/Software – Event management (triggers/alerts) – Query & reporting tools – Enterprise BI platforms – Heat maps/dashboards/ balanced scorecard tools reflecting risk – Predictive analytics/process modeling tools for measuring and monitoring risk

Aberdeen Group CPO Survey Report & SCRM Report 2010

FIGURE 2.2
Best-in-class SCRM framework.

risk management arena. The bulk of Aberdeen's findings about and journey toward good SCRM hold true today.

Also in 2010 PRTM (now part of PricewaterhouseCoopers) released its *Global Supply Chain Trends Report.*[4] A few of the novel themes emanating from this study were that almost 75% of all the respondents believed that demand and supply volatility will become the biggest roadblock to capturing profits in the economic upturn, and more than 80% expected complexity in their supply chains to increase dramatically, partly due to an increase in product proliferation. The PRTM study produced several important takeaways:

- Demand and supply volatility is here to stay.
- Supply chain complexity will be with us for many years.
- End-to-end supply chain cost optimization will be critical moving forward.
- Risk and opportunity management should span the entire supply chain, including with key partners.

This research concluded, and this has certainly been borne out by subsequent experience, that supply chains will become more volatile and complex. The researchers also concluded that most companies are ill-prepared to manage this complexity and that risk management will be a critical success factor affecting corporate success.

We also saw two more reports explicitly targeting supply chain disruptions and their financial impacts. Chainlink Research and Hendrick & Singhai had been working on attempting to codify the financial impacts of supply chain disruptions on company bottom lines. According to these resources, a publicly traded company that experiences a moderate to severe supply disruption should expect to realize a 107% drop in operating income on average, 114% drop in return on sales, 93% drop in return on assets, 7% lower sales growth, 11% growth in cost, 14% growth in inventories, and a 10% reduction in shareholder value. This is one of the first reports that linked risk events to tangible performance shortfalls.

Finally, a report from the Business Continuity Institute (BCI) in the United Kingdom revealed serious levels of supply chain failures and disruptions around the globe. This study found that weather disruptions affected more than half of all companies in 2010, up from 29% the previous year. Unplanned IT and telecommunication outages were second on the list of most frequent disruptions. More than one third of companies mentioned service failures by outsourcer providers, up from 20% the previous year. More than 50% of the respondents stated that their disruptions led to a loss of productivity and 20% of the respondents admitted they had suffered damage to their brand or reputation as a result of a disruption.

The report concluded that the serious levels of supply chain disruption experienced by organizations around the globe, coupled with the wide range of threats, underscores the business case for investment in business continuity planning (BCP). We will go into more detail later about BCP. For now, consider BCP to be an internal insurance policy or risk response plan that identifies, assesses, mitigates, and manages risk scenarios.

2011

During 2011 Accenture began to talk about the "new normal" of global supply chains. The new normal featured more global sourcing and manufacturing, hyper-demand requirements, longer supply chain lead times, more potential points of failure, and managing supply chains in far-flung corners of the world.[5] With these elements of supply chain risk coming to the surface, Accenture produced a set of principles that support the birth of supply chain risk management as a discipline. These six principles include the following:

1. Integrate risk management practices across all business functions to ensure understanding, commitment, and alignment
2. Identify, measure, and prioritize risks by mapping out the complete supply chain "ecosystem"
3. Emphasize operational flexibility, global visibility, and diversified supplier portfolio to blunt the impact of supply chain calamities
4. Use probability modeling to identify unknown risks and develop contingency plans
5. Insist that suppliers and business partners perform up-front due diligence
6. Hedge risk by making prudent choices about insurance

We will dig deeper into every one of the six tenets throughout the book.

AMR surveyed more than 500 executives worldwide during 2011 and asked questions about supply chain risks, disruptions, effects of those disruptions, and methods to mitigate and manage risk. Supply failures were the most cited disruption noted by executives. Quality failures, natural disasters, and commodity price volatility were also cited by almost one third of respondents. In terms of the most common risk management approaches, executives said they relied on (in order of mention) meetings/discussions with external partners and suppliers, predictive analytic tools, staying current with supplier business continuity plans, utilizing reports/data from third-party sources, and utilizing performance risk dashboards.

In October 2011 SCM World, the global institute for supply chain learning, training, and development at Stanford University, released its annual *Chief Supply Chain Officer* (CSCO) study.[6] The survey involved 750 global executives of which over 50% were vice presidents or higher. The report was conducted during the immediate aftermath of the Japanese earthquake and tsunami, creating an interest in understanding how supply chains were reacting to these disruptions. The report highlighted the disruptions due to the tragic events, how many companies had suffered shortages in supply and capacity, lost sales, inventory write-offs, and plant closures across many industries. Even though the report didn't examine supply chain risk specifically, it did talk at length about how the supply chain is a critical success factor and a driving force for competitive advantage.

Another key theme throughout the report was that supply chain sustainability is critical in terms of driving value for the corporation, especially in times of demand and supply volatility.

During this period we also picked up on definitions of a resilient supply chain from the Supply Chain Council.[7] While the Supply Chain Council was updating its SCOR Model and risk definitions and frameworks, the importance of resiliency in the supply chain came to the surface. Essentially, the council concluded that in a world of technical change, financial risk, political turbulence, and mounting regulatory pressures, industry growth does not always proceed smoothly. Risk management is especially challenging when threats are unpredictable. At the same time, corporations are accepting broader responsibility for social and environmental impacts of their supply chains. A resilient enterprise has the capacity to overcome disruptions and continually transform itself to meet the changing needs and expectations of its customers, shareholders, and other stakeholders.

2012

Another study by the Business Continuity Institute addressed the impact of the Great East Japan earthquake and Christchurch earthquake in New Zealand. The key take-away from this research was the wide-ranging differences in an important SCRM metric, time-to-recovery. BCI assessed how long it took supply chains to recover from the impact of the earthquakes. The compelling statistic is that almost 50% of all the respondents needed more than five weeks to recover.

Aberdeen Group released a report on risk from a financial perspective.[8] Coming off the heels of the Japanese disasters, the report attempted to understand the risk-adjusted strategies actually operating inside the CFO's office. The report concluded that whether a company is looking to reroute a supply source away from a natural disaster or seeking to mitigate the risk associated with tax audit exposure, it is clear that companies with processes and tools that enable clear visibility to risk entities and a means to react quickly are going to be the ones that hold the key to effective budgeting and profitability.

Zurich, one of the largest insurers in the world, published a set of supply chain risk statistics and some revealing numbers associated with supply chain business interruptions.[9] The insurer released a report that profiled the causes behind supply chain disruptions derived from an analysis of insurance claims over a several-year period. Clearly, this analysis addressed primarily hazard risk. The insurer found that 85% of organizations experienced at least one supply chain incident that caused disruption to their business. More than 50% of supply chain disruptions occur because of adverse weather, and just over 40% occur because of unplanned IT or telecom outages. Other significant disruptions occurred because of loss of talent/skills, product quality incidents, civil unrest/conflicts, and cyberattacks. Of the disruptions that occurred, nearly 40% originated below the tier-one supplier level.

A research article titled *Researcher's Perspectives on Supply Chain Risk Management* presented a study of the diversity of perspectives surrounding supply chain risk management.[10] This study identified three gaps in the body of knowledge:

- **Definition gap**—there is no clear consensus on the definition of SCRM because many limit the scope of SCRM to rare but large events, while others believe that SCRM is about supply–demand uncertainties. (Hopefully, the definitions we presented in Chapter 1 help to clarify this term.)
- **Process gap**—there is a lack of research on an important aspect of the risk management process, namely, the response to supply chain risk incidents.
- **Methodology gap**—there is a shortage of empirical research in the area of SCRM.

During this period an important study on supply chain risk was concluded by Deloitte and Forbes Insights. This survey covered 192 U.S. executives, CEOs, CFOs, SVPs, and directors across multiple industries. One finding that stands out from this work is that 91% of respondents said they planned to reorganize and reprioritize their approaches to risk management during the subsequent three years.

A disconcerting aspect of this survey was the completely scattered approach toward risk responsibility. Fully one quarter of respondents indicated the CEO is primarily accountable for risk management; almost a quarter said the said the CFO/treasurer's office is primarily responsible; almost 20% said the chief risk officer/treasurer is responsible; and fewer than 15% said legal/compliance is responsible. The remaining respondents were spread across various other groups. Future risk management plans include more than half of respondents saying their company planned to elevate risk management within their organization, while almost 40% said they will reorganize to support enterprise risk management (ERM), a framework introduced in Chapter 1. Almost 40% of respondents said they will provide more staff training, while almost one third planned to incorporate more technology into their risk management efforts.

And finally, SCM World, which we referenced previously, released its annual *Chief Supply Chain Officer* report. This report surveyed more than twice the number of companies as in 2011 and explored five main topics, one of which was risk management. A novel aspect of this survey involved SCM World asking respondents what they were doing to identify, assess, mitigate, and manage risk and profiled the impact of risk events from both demand and supply disruptions. From this study we learned that the highest impact from supply and demand disruptions over a two-year period was a loss of sales/revenue. In order of impact, subsequent impacts included lower profits, delays in product launches and growth plans, loss of customers, higher cost of capital, damage to image and reputation, and lower share price/shareholder value. This research revealed the many damaging effects of supply and demand disruptions.

2013

The start of 2013 featured the release of a report from the World Economic Forum (WEF) in Davos, Switzerland. The sponsors of the report were Zurich Insurance, Accenture, Partners against Corruption (PACI), and the World Economic Forum. The report, titled *Building Resilience in Supply Chains*, was an outcome of WEF's initial Supply Chain Risk Initiative started in 2011. A clear finding in this report is the need for organizations to shift from reactive to proactive risk management. Another relevant finding is that more than 80% of companies are now concerned about supply chain resilience. Study participants also indicated there is a need for a common risk vocabulary and that cyber risk may have the greatest risk implications for supply chains.

The World Economic Forum's report touched on what it will take to make SCRM a bona fide business discipline. Perhaps most importantly, risk management must become an explicit and integral part of supply chain governance. Other high-level suggestions for moving SCRM to a business discipline include the following:

- Institute a multistakeholder supply chain risk assessment process across the enterprise
- Mobilize international standards' bodies to further develop resilience standards
- Incentivize corporations to follow agile, adaptable supply chain strategies
- Expand the use of data-sharing platforms for risk identification and responses

A RIMS.org article mentioned the Risk Maturity Index developed by AON Insurance and the Wharton School of Business. The index suggests that companies with the highest level of risk maturity (a measure that gauges the development of an organization's risk strategy and framework) experience 50% lower stock price volatility than less-developed counterparts. Over a two-year period (2010–2012) companies with higher risk maturity ratings saw greater annual stock price returns. This was especially apparent when the only companies to see positive returns during that volatile period were those with the highest risk maturity levels. Those lower on the maturity scale saw losses between 17% and 30%. This work again pointed out the critical relationship between risk management and financial outcomes.

Lloyd's of London released several statistics in 2013 associated with the Japanese earthquake and tsunami and Thailand floods. The first statistic they shared was the combined property and business interruption losses, which reached a record-breaking $240 billion, with just $47 billion of the losses covered by insurance. The insurance industry provides supply chain interruption products called contingent business interruption (CBI). However, a large majority of companies impacted only had asset-based property damage insurance. Some of the elements covered by CBI include getting workers in and out of damaged facilities, working to get power to facilities, and shipping and receiving goods into and out of facilities. Chapter 5 will reexamine the floods in Thailand from a risk quantification perspective.

An interesting report from the Association of Insurance and Risk Managers in Industry and Commerce (AIRMIC) titled *Supply Chain Failures: A Study of the Nature, Causes, and Complexity of Supply Chain Disruptions*, identified seven underlying factors that tend to be present whenever supply chains go wrong: off-shoring, increasing complexity, cost pressures, geographic clustering, modern communications, modern production methods, and increasing dependency. The report also estimated that economic losses from supply chain disruptions have increased 465% between 2009 and 2011.

And finally, Ernst & Young interviewed more than 420 CFOs and heads of supply chains at technology, automotive, manufacturing, aerospace, and defense companies. The report concluded that when CFOs and supply chain leaders form a closer business partnership within a company, they report better results in a number of areas, including the company's financial position. The study found, for example, that 70% of CFOs and 63% of supply chain executives said their relationship had become more collaborative over the past three years. The merger between finance and supply and supply chain professionals appears to be an inevitable one.

The intriguing outcome from this survey is, because CFOs take a long-term approach to formulating business strategy, they are in a unique position to manage risk and plan for business continuity, something that challenges supply chain managers who tend to think in terms of shorter time horizons. The CFO also has the opportunity to work with the procurement and treasury groups to determine the extent to which risk is owned and managed by the company and the extent to which it is pushed down through the supply chain.

In 2013 the authors of this book published several articles on SCRM. One key article was *SCRM: The New Discipline of Supply Chain Excellence*. After several years of teaching in the classroom and conducting workshops around the globe, we became thoroughly convinced of the need for SCRM to become a discipline. The article identified what was emerging from the classroom and from the practical application of the body of knowledge in the field. This article was a major impetus for this book.

FOUR PILLARS OF SUPPLY CHAIN RISK MANAGEMENT

We would like to finish up our "as-is" discussion by providing an assessment of the state of supply chain risk management. This will involve something we call the Four Pillars of SCRM. These pillars include supply risk, process risk, demand risk, and environmental risk. Each pillar encompasses its own set of tools, techniques, tactics, metrics, people, processes, and program issues. The complexion of each pillar is below.

Supply Risk

The complexion of this pillar encompasses areas such as supplier continuity, strategic sourcing, supplier viability and capability, raw material pricing, supplier assessments, inbound logistics, fraud, corruption, and counterfeiting. Inherent risks here are disruptions caused by the inability of suppliers to deliver on time, quality failure, financial failure, compliance failure, channel complexity, and communication failure.

Process Risk

This pillar includes IT systems, mergers and acquisitions, marketing strategy, organizational structure, frameworks and metrics, supply chain strategy and execution, manufacturing and quality, organizational risk assessment, heat maps, and war rooms. The inherent risks here include disruptions caused by quality problems, inventory shortages, late deliveries, capacity shortages, equipment breakdowns, IT outages, poor overall execution, and misalignment of strategy and metrics.

Demand Risk

The complexion of this pillar covers areas such as new customers, market trends, consumer interest/spending, demand management/forecasting, distribution requirements planning, product integrity, customer service, and scenario planning. Inherent risks here are disruptions caused by problems in distribution, actions by competitors, product reputation, brand management, social media/trending, logistics, and customer sentiment.

Environmental Risk

This final pillar encompasses areas such as government regulations, taxes, economic volatility, currency exchange, natural disasters, and compliance. Inherent risks are natural disasters, geopolitical and energy risks, port security, logistics and facilities security, currency exchange fluctuations, global economics, war, pandemics, and civil disobedience.

We have compiled a profile of the maturity and activity level for each pillar. Figure 2.3 depicts our assessment of the maturity and activity level

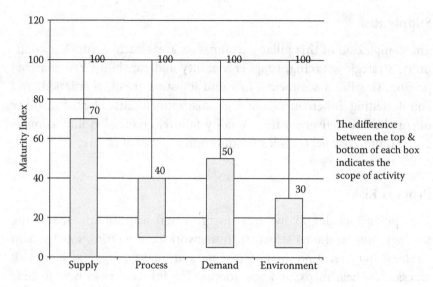

FIGURE 2.3

Four-pillar SCRM maturity and activity level.

for each pillar. On the left or y-axis is the maturity level of each pillar. The horizontal x-axis depicts the Four Pillars, and the size or length of each "box" is an indication of the activity level within each pillar.

From a maturity point of view, we feel the supply pillar is by far the most mature of the four, positioned at about 70 out of 100. Why? Procurement professionals have been dealing with supplier uncertainty and risk for more than 50 years. This discipline has become a profession, supported by member-driven organizations who are providing certifications to demonstrate that these professionals have a command of the body of knowledge and best practices. Tools such as supplier relationship management (SRM), spend management, credit and financial reporting by public credit organizations, and more have matured over the years. And the present activity level of new techniques such as supplier risk assessment; supply chain mapping; and fraud, bribery, and corruption identification supported by new cloud-based software systems is providing the procurement professionals with a host of new tools and techniques to leverage in an effort to identify, assess, mitigate, and manage supplier risk.

The next-highest level of maturity and activity, in our opinion, is the demand pillar. Positioned at about 50 out of 100, demand management solutions, such as sales forecasting and many other deterministic tools have been around as long as the supply tools. So why then do we feel this pillar is at a lower maturity level than supply? The tools utilized in this pillar have been developed to support processes that are purely forward looking and one-dimensional, such as sales forecasting, and do not involve any element of uncertainty or risk. Another tool, collaborative planning, forecasting, and replenishment (CPFR), does provide some coverage of risk because it attempts to support information sharing between supplier and customer in an effort to minimize demand surprises and shocks to the supply chain. New techniques and tools such as probabilistic planning, discrete-event simulation, and digital modeling are emerging to assist demand managers and sales and operations planning (S&OP) process owners to run "what-if" scenarios that will demonstrate how their supply chains will act when a risk event shocks their organization. These new tools overtly handle uncertainty and risk and will take some time to mature.

Next in the maturity and activity level is the process pillar. A tremendous number of tools and techniques support all the processes we've highlighted in this pillar. Many have been around for more than 40 years and are supported by professional organizations such as APICS, CSCMP, ASQC, ISSSP, and others and also include professional certifications. Again, our reasoning for the pillar's positioning is that many of the tools do not embrace uncertainty and risk. They support discrete, linear functions, such as planning inventory, planning capacity, production scheduling, quality, logistics, and more. These functions are driven by solid supply chain management metrics such as maximizing service, reducing cost, and improving asset utilization, not mitigating risk.

And finally, the environmental pillar is very new and continues to expand because of new industry-specific and governmental rules and regulations. There is more and more activity in this area, but with new and ever-changing regulations, this pillar will take a long time to solidify.

THE SUPPLY CHAIN RISK MANAGEMENT ADOPTION

We'll finish this chapter with a brief discussion on what we call SCRM Adoption. This utilizes a categorization scheme revolving around laggards, industry average, and early adopters. We want to leave you with a sense of the operational complexion for each category in terms of SCRM along with a few salient statistics from the early adopter companies.

SCRM Adoption

We know that leading-edge companies are much more likely to integrate and align risk with corporate goals. They generally have more visibility into their organization as they are substantially more likely to perform "what-if" scenario-planning and change analysis. On average, leaders achieve almost 95% better accuracy of cash flow forecasts, which is clearly better than what their peers achieve. And they tend to mitigate their financial losses down to 3% of revenue as opposed to their peers, who average 10% of revenue due to financial loss. What characterizes early adopters, industry average, and laggards?

Early Adopter Companies. Early adopters leverage their ERM tools and technology to enhance the integration of risk management across the business. They continually link risk management to company goals and compensation. And, they are improving visibility and monitoring of Key Risk Indicators with new business analytic tools.

Early adopters also implement processes that are aligned with risk management and compliance, build a risk awareness culture throughout the organization, and secure executive commitment for risk management initiatives. These practices are being driven by the early adopters by a factor of two- or three-to-one above the industry average and laggard companies.

Other attributes characterize early adopters. These companies maintain a senior management champion of risk, segment risk duties, cross-functionally coordinate risk management, establish roles and responsibilities to execute risk, and establish a risk committee to oversee key risks. These commitments by early adopters eclipse the industry average and laggards again by a factor of two- or three-to-one.

Industry Average Companies. Average companies attempt a standardized approach to communication and organizational collaboration relative to risk. Some of these companies are beginning to integrate and align risk with corporate goals. Some are also developing and measuring risk and performance. Several are beginning to improve the time-to-decision by optimizing risk knowledge management activities and expanding risk visibility throughout the organization. Senior management oversight and engagement associated with risk to improve executive buy-in is starting to occur at this level.

Laggard Companies. Laggards talk about supply chain risk but do not fund projects to prepare for and respond to risk events at any real level. Most have someone in the CFO's office reviewing enterprise risk in the

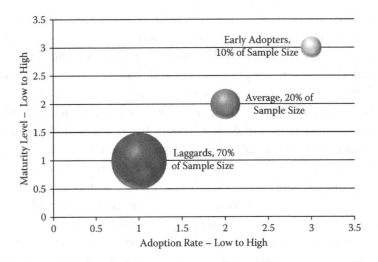

FIGURE 2.4
SCRM adoption.

classical financial terms, such as hazard, financial, and strategic risk. The average operational and supply chain professional does not maintain scenario game plans for risk events; therefore, SCRM is an event-driven, ad hoc, or part-time experience. Most executives in this category assume incorrectly that their people will know what to do in a risk event.

Figure 2.4 depicts a graph encompassing laggards, industry average, and early adopter companies in terms of the adoption rate of SCRM concepts. In terms of the as-is state, it is safe to conclude that a majority of companies are still considered laggards in terms of their SCRM capabilities. This suggests that we have more work to do to move companies away from the laggard status. Subsequent chapters will concentrate on how to do that.

CONCLUDING THOUGHTS

The purpose of this chapter is to help us understand the current state of supply chain risk management. From our analysis we can reach some overarching conclusions. First, the negative outcomes from supply chain risk events are often quite severe. Only a fool would believe otherwise. Second, no consensus exists among companies or industries concerning how to organize for or manage supply chain risks. The ways that companies approach SCRM are varied. Third, most companies are ill-prepared from an employee, measurement, organizational, and IT perspective to operate in an environment characterized by increased uncertainty and volatility.

Another major conclusion is that most companies recognize the threat of supply chain uncertainty and understand that operating in a condition called the "new normal" will not be as enjoyable as operating in the "old normal." A study by Zurich Insurance, for example, reported that 75% of respondents state they still do not have full visibility into their supplier base. This is concerning because several studies have concluded that supply disruptions are the most widely cited supply chain risk event.

Finally, most SCRM efforts rely on heroics rather than planning and prevention. In terms of SCRM maturity, most companies are considered laggards with some movement toward average. Minimal proactive supply chain risk management appears to be occurring. Growth along the maturity curve needs to accelerate, especially when we consider that almost 75% of risk managers believe that supply chain risk levels are higher than just a few years ago and that risk will continue to increase. More than 70% of risk managers say the financial impact of supply chain disruptions has also increased compared with just a few years ago.[11] As a sign of the times, the Allianz Risk Barometer in 2013 ranked for the first time ever business interruption and supply chain risks as the top concerns of businesses globally. One of the main objectives of this book is to prepare us to manage in this "new (ab)normal."

Summary of Key Points

- Most supply chain executives became interested in SCRM after the financial meltdown of 2008. For many, 2008 was the genesis of their risk management efforts.
- In 2009, the ISO Group delivered its first set of standards directly relating to supply chain risk, which was a major recognition of the importance of SCRM. These standards include ISO 73 and ISO 31000.
- The Supply Chain Risk Assessment Tool encompasses about 100 questions-of-discovery about a company's supply chain across 10 tenets covering the entire supply chain. The basic premise is that as the supply chain matures, the inherent risks faced by that supply chain diminish.
- The new normal of global supply chains features more global sourcing and manufacturing, hyper-demand requirements, longer supply chain lead times, more potential points of failure, and managing supply chains in far-flung corners of the world.

- Over a period of several years, Zurich found that 85% of organizations experienced at least one supply chain incident that caused disruption to their business.
- The Risk Maturity Index suggests that companies with the highest level of risk maturity experience 50% lower stock price volatility than less-developed counterparts.
- The Four Pillars of SCRM are supply risk, process risk, demand risk, and environmental risk. The supply pillar is by far the most mature of the four because procurement professionals have been dealing with supplier uncertainty and risk for over 50 years. The demand pillar ranks second in terms of maturity, followed by the process pillar and finally the environmental pillar.
- SCRM Adoption utilizes a categorization scheme revolving around laggards, industry average, and early adopters. In terms of the as-is state, it is safe to conclude that a majority of companies are still considered laggards in terms of their SCRM capabilities.

ENDNOTES

1. "Managing Risk for High Performance in Extraordinary Times." *Accenture 2009 Global Management Study.* 2009.
2. Arntzen, Dr. Bruce, Prof. Maria Jesus Saenz, and Isabel Agudelo. "The SCALE, Supply Chain & Logistics Excellence Network." *MIT's Global Scale Risk Initiative,* MIT's Center for Transportation & Logistics, March 2010.
3. Accessed from Aberdeen Group CPO Survey Report & SCRM Report, 2010.
4. Burson, Patrick. "PRTM's Global Supply Chain Trends 2010–2012 Survey." *Supply Chain Management Review,* (June 2010).
5. Pearson, Mark. "Inoculate against Supply Chain Risk." *Logistics Management,* April 2011: 20–21.
6. Lee, Hau, PhD, and Kevin O'Marah. "Chief Supply Chain Officer Report." *SCM World,* October 2011.
7. Pettit, Timothy J., Joseph Fiskel, PhD, and Keely L. Croxton, PhD. "Can You Measure Your Supply Chain Resilience?" *Supply Chain and Logistics Journal, Canada,* Spring 2008.
8. Catellina, Nick, and William Jan. "Leveraging Risk-adjusted Strategies to Enable Corporate Accuracy." Adapted from Aberdeen Group, April 2012.
9. Accessed from Zurich Re's Knowledge Vault on Risk, July 2012.
10. Sodhi, Manmohan S., Byung-Gak Son, and Christopher S. Tang. "Researcher's Perspective on Supply Chain Risk Management." *Production and Operations Management,* (2011), Accessed from Jan Husdal's SCRM blog (www.husdal.com).
11. Favre, Donovan, and John McCreery, "Coming to Grips with Supplier Risk." *Supply Chain Management Review,* 12, 6 (September 2008): 26. Citing statistics from Marsh, Inc. and *Risk & Insurance* magazine.

3

Building the Risk Management Foundation

Achieving excellence in any area does not happen because a company simply announces its desire to be excellent. Many organizations have been frustrated because they lack the ability to develop the kinds of supply chain approaches and techniques that create a differential advantage. These organizations fail to recognize the importance of mastering something we call the "enablers," four distinct areas that support more advanced supply chain and risk management initiatives.

This chapter highlights a set of enablers that are essential for effective risk management. These enablers include a supportive organizational design; information technology systems that provide real-time or near real-time access to data and information; risk-related measures and measurement systems that provide insight into potential risks as well as the effectiveness of risk management efforts; and the availability of capable human resources. This chapter also discusses how to integrate risk management with supply chain strategy development. We conclude with a case that describes how one organization relied on these four enablers, as well as a centrally led approach to strategy development, to manage strategic risk.

SUPPLY CHAIN RISK MANAGEMENT ENABLERS

As mentioned, what separates companies that achieve real advantage from their risk management efforts is a commitment to the four enablers of supply chain excellence. While we could commit entire chapters to each enabler, the following provides an overview that should help the reader understand what we are trying to present. If 50 books are written next year addressing risk management, the probability that any of them commits any space to these four enablers presented here is slim. And, in our opinion, that is a serious mistake.

A Supportive Organizational Design

Perhaps one of the most underappreciated parts of corporate success is the role that organizational design plays. The Corporate Executive Board has concluded that to be successful, executives must first consider how their organizational design can lead to sustainable improvements in performance and operational excellence. Organizational design refers to *the process of assessing and selecting the structure and formal system of communication, division of labor, coordination, control, authority, and responsibility required to achieve organizational goals.*[1] An organization's design, including the varied features put in place to support that design, is much more than what an organizational chart can ever depict. Because this topic does not appear elsewhere in the book, it will receive the most attention of the four enablers.

In our experience, supply chain risk management (SCRM) is rarely established as a distinct function within a company, although many organizations have a "point person" who has responsibility for risk management. One conclusion that we (as well as others) have reached is that businesses have not yet agreed upon any typical way to integrate SCRM into their decision-making processes.[2] The challenge is not about creating an organizational design that is dedicated to risk management, something that is unlikely to happen at most organizations. Rather, the challenge is to take various design features, many of which are already in place, and incorporate risk management responsibilities. The following illustrates a variety of design features that are ideal for including risk management as part of their scope.

S&OP Processes. Sales and operations planning is an internal, cross-functional process, usually supported by teams, whose primary output is a 6- to 18-month production schedule for product categories and families. The objective of S&OP is to develop an output plan that minimizes total costs given a specific demand plan. This process is an ideal way to bring the demand and supply sides of a value chain together, something that in itself has positive risk implications. Given that cost minimization is a major objective of any S&OP process, finance will play a major role in any planning exercises. The S&OP process, which is conceptually simi-lar to the more traditional aggregate planning process, formally reviews customer demand and supply resources and updates plans quantitatively across a rolling time horizon. An S&OP process may be one of the best preventive steps a company can take to address operational risk, particu-larly those risks that arise from poor supply and demand planning.

Companies have been creative in how they practice S&OP. A central supply chain planning group at a leading chemical company has extended its S&OP concept by assuming responsibility for all of the activities associ-ated with demand and supply planning and execution except production. Using sophisticated algorithms, supply chain planners have responsibility for managing the flow of a product's raw materials and information from suppliers all the way through to customers. No hand-offs of information take place between demand and supply groups, creating a clear respon-sibility and accountability for demand planning, supply planning, and customer service. Other companies have expanded their S&OP process to directly include inventory planning and management, calling their process S&IOP. Here, the letter "I" represents inventory planning. Advanced com-panies are also beginning to model external risk probabilities and scenar-ios into their S&OP process. Chapter 14 will highlight one such company.

Collaborative Planning, Forecasting, and Replenishment (CPFR) Process. CPFR follows a defined organizational framework that combines the intelligence of multiple trading partners in the planning and fulfill-ment of customer demand across a supply chain. It has the stated objective of increasing product availability to customers while reducing inventory, transportation, and logistics costs.[3] A key part of CPFR involves collab-orative forecasting, which is the process of collecting and reconciling information within and outside the organization to come up with a single projection of demand. Like S&OP, the direct linkages between effective planning and better risk management should be clear. And, like S&OP, it is relatively easy to see how this process can be expanded to consider risk issues during the planning process.

Executive Responsibility. Another organizational approach involves making a single executive ultimately responsible for demand and supply planning activities. A leading U.S. company has created the position of vice president of supply chain management and charged that executive with responsibility for worldwide supply planning and replenishment, demand and finished good forecasting, inventory planning, primary customer order fulfillment and logistics, and integrating supply chain

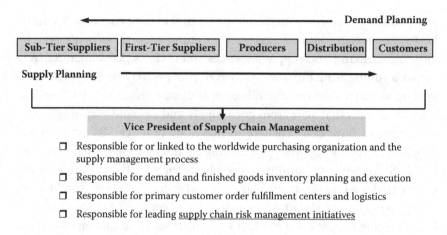

FIGURE 3.1
Executive supply chain position.

activities with operational positions. The primary objective here is to create a single point of accountability for satisfying end customer requirements at the lowest total cost. A logical step would be to expand these responsibilities to include risk planning and assessment. Figure 3.1 illustrates how an executive position that also has oversight for supply chain risk management might be structured.

Aligned with the idea of executive responsibility (and accountability) is the position of a chief risk officer. No clear agreement exists among academics and practitioners whether a chief risk officer responsible for supply chain risk is a good idea, although a survey by Accenture reveals that at the corporate (i.e., enterprise resource management, ERM) level, 98% of organizations have a chief risk officer.[4] The need for a chief risk officer within the supply chain domain is often situation specific and will vary from company to company, perhaps even from industry to industry. It is always tempting to create a new position, whether it is the czar of lean, total quality, or sustainability, and to charge that individual with responsibility for "making things happen." Isn't this new position visible evidence that decisive action is being taken? If only it were that easy. The success record for such positions is, unfortunately, mixed. These positions often lack the resources and authority to effect meaningful change across well-protected organizational turfs. With this model, far too many people are content with letting someone else take responsibility for managing risk, and conversely, taking the blame when things do not go well.

Organizing around Processes. Something that we know for certain is that effective risk management requires open sharing of supply chain information. Another thing we are certain about is that functional organizational designs, while generally effective at moving information up and down the command and control structure, are not nearly as effective sharing information laterally across functional groups. Unfortunately, as it relates to supply chains, vertical movement increases the possibility that the right groups will not have access to the right kinds of information, including risk information. Functional designs tend to be risk management inhibitors rather than enablers. We can safely assume that effective risk management will benefit from a horizontal (i.e., cross-functional) rather than vertical (i.e., functional) view of the supply chain.

One way to counter any potential issues or risk from inadequate sharing of information or poor coordination of activities across the supply chain is to create a process-centric organization. Numerous books and articles have been written highlighting the advantages of a process-centric organization.[5] Research evidence suggests that an often-predicted movement away from strict functional alignments is already occurring. One study revealed a clear link between organizations that are structured around major processes and their ability to attain their supply chain objectives, including better management of risk.[6] This study also concluded that process-centered design features, such as the use of cross-functional teams along with executive positions responsible for overseeing processes rather than narrower functional tasks, should show large increases in usage compared with more traditional design features.

Taking a process view helps manage the conflicts and trade-offs that inevitably occur as work crosses functional boundaries. Ineffectively managing these conflicts and trade-offs can have serious risk implications. (A trade-off is a balancing of factors, all of which are not attainable simultaneously.) Left unattended these crossing points can easily lead to conflict, competition, and inefficiency. A process orientation, at least in theory, should promote the seamless movement of work and information across boundaries.

Part of the reason that conceiving of work in terms of processes can be challenging is that most of us have been trained to think in terms of physical processes. This includes the technology and equipment to convert raw and direct materials into finished products. These are the processes that we see, touch, and hear. But that is not the only kind of process that we see in supply chains. Most of us work within organizational or business (i.e., nonphysical) processes such as fulfilling customer orders, developing new products or services, estimating customer demand, or evaluating and selecting suppliers. Some refer to organizational processes as business processes.

Organizational Work Teams. Whether we like it or not, the use of teams is just about everywhere in supply chains and is a popular and growing organizational design option. A partial listing of teams and groups involved with some aspect of supply chain management is as follows:

- Customer Advisory Board—An executive-level group that brings suppliers, customers, and the original equipment manufacturer (OEM)/producer together to share information such as end customer requirements and risk issues
- Buyer–Supplier Council—An executive-level group that includes an OEM/producer and a rotating group of key suppliers to share information such as product and strategy development plans
- Global Sourcing Steering Committee—An executive-level group that has responsibility for overseeing centrally led global sourcing initiatives
- Commodity Management Team—A team that has responsibility for managing commodity groups or categories of purchase requirements with responsibility for supplier selection decisions, relationship management, and increasingly risk management
- Demand and Supply Planning Team—A cross-functional planning group that identifies and balances downstream customer demand requirements with upstream supply availability
- Buyer–Seller Improvement Team—A cross-organizational team that focuses on tangible supply chain improvement opportunities and projects between the buyer and seller, such as supplier development opportunities
- Value Analysis/Value Engineering Team—A continuous improvement team that has responsibility for enhancing customer value by systematically analyzing the relationship between product/service function and cost
- Customer Order Fulfillment Team—A cross-functional team that has responsibility for managing the customer order fulfillment process from order entry to final delivery, sometimes including through final customer payment

- New Product Development Team—A cross-functional team that has responsibility for developing new products and services, ideally with supplier and customer support
- Supply Chain Process Management Team—A cross-functional team that has responsibility, usually on a full-time basis, for managing and improving critical supply chain processes
- Customer Relationship Management and Service Teams—Teams that have responsibility for managing longer-term relationships with key customers

What does such a wide variety of groups and teams have to do with supply chain risk management? While the use of teams presents challenges that are beyond the scope of this chapter, they also present a tremendous opportunity to embed risk considerations into many settings. Every team listed above should address risk issues as part of their scope.

The use of organizational design features across a supply chain is varied, which means that opportunities exist to engage in risk-related discussions that consider supply chain management and risk management topics simultaneously.

Information Technology

Information technology supports risk management excellence by enabling the seamless availability, analysis, and movement of supply chain data and information, thereby allowing more informed decisions. Regardless of the type of information technology platform or software used, supply chain systems should have the ability to capture and share information across functional groups and organizational boundaries in real time or near real time. This may involve transmitting the location of transportation vehicles with global positioning systems, transmitting material requirements with web-based electronic data interchange systems, or capturing demand and replenishment data using bar code technology. Radio frequency identification (RFID) tags are also beginning to capture real-time data about material and product movement.

It is important to understand the important role that shared, real-time data can provide to risk managers. A powerful outcome occurs when real-time data and predictive analytics are combined to anticipate what will happen rather than react to what did happened. Knowing what will happen with a relative degree of certainty enables supply chain mangers to take preventive action, something that is the holy grail of risk management.

There is no question that "big data" is the next big thing in the IT world. Big data is the efficient analyzing of massive amounts of the data that organizations collect every day. Combining massive amounts of data with sophisticated algorithms comes under a variety of names, including big data, predictive analytics, data analytics, and business analytics. Predictive analytics are being used to anticipate customer purchases, identify demand patterns, predict the occurrence of crimes (something called predictive policing), anticipate patient heart attacks, and identify truck maintenance problems (to name but a few applications). Its relevance to better risk management practices is obvious.

Some companies scan massive amounts of data to determine the best combination of operating variables to achieve a desired outcome at the lowest total cost. A major trucking company, for example, monitors in real time every aspect of the performance of its fleet as well as the driving behavior of its drivers. Is a driver driving too fast or slow? Is the engine performing properly? Is the driver letting the truck idle too long? What is the best combination of operating parameters that yields the highest miles per gallon? Besides helping the company optimize its fleet operations, the data also track compliance to hours of service regulations. Of course, the drivers have their own opinion about this real-time visibility.

What has made the predictive analytics suddenly become popular is the development of Hadoop (named after the developer's son's toy elephant), an open source system that mimics the "divide and conquer" approach taken by Google.[7] This system can also handle data that do not fit neatly into a spreadsheet. The Hadoop approach spreads tremendous amounts of information across data centers and then probes that pool of data quickly and inexpensively for answers to queries. This is in contrast to traditional data warehouses that cram as much well-organized information as possible onto a few expensive computers and then "crunch" the data as the system searches for answers. A second wave of new start-ups, as well as some big-name companies, is developing easier-to-use versions of Hadoop. They are also extending the use of inexpensive servers to analyze new categories of data, including blogs, pictures, videos, tweets, and even medical images. The ability to use analytics to predict risk events will be a "game changer." This topic is so important that Chapter 11 focuses on using analytics to predict the future and manage risk.

Measurement Systems

A third enabler includes the measures and measurement systems that are in place to provide insight into the many facets of SCRM. Measurement systems in general come under criticism for some good reasons (too many measures, too few measures, measures that drive the wrong behavior, measures that are backward looking, measures that focus on activity rather than accomplishment). In the risk management arena, the most obvious criticism is the lack of measures that relate directly to risk, although many measures have indirect risk connections. A parts-per-million quality metric provides insight into quality problems, which will have a secondary relationship to operational risk. What we often fail to see in supply chain management are measures whose primary objective is to capture and report on some element of supply chain risk. Chapter 13 will present a set of leading-edge measures that address supply chain risk specifically.

Why do we even need risk-related measures and measurement systems? First, objective measurement supports making fact-based decisions about risk issues. Measurement is also an ideal way to communicate information across a supply chain and to promote continuous improvement and change. Measurement also conveys what is important by linking critical risk-related measures to strategies and desired business outcomes. Finally, measurement helps identify whether new initiatives are producing the desired results. Measurement may be the single-best tool to make risk management an embedded part of a company's culture.

An example of a system that uses defined measures and algorithms to focus directly on risk is Boeing's SINC (Supplier Insight and Control) system. The vision of the SINC process is to deliver early identification of supply chain risks and improve decision-making capabilities. In this system, suppliers input data that pass through various measures and algorithms to arrive at performance predictions. The SINC process promotes collaboration with suppliers and provides early risk identification through the collection and use of Boeing and supplier process data. The SINC system focuses on insight rather than oversight.

While measurement is a critical enabler, it is certainly not unique to supply chain or risk management. A convincing argument can be made that risk measurement systems are not nearly as developed as they are in finance or marketing, for example. It might be tempting to ask why measurement is important enough to be considered a risk management enabler. The bottom line is that any major strategy or change, including cultural changes, requires a way to validate success. Measurement is a primary way to validate that success.

Talent Management

Something that we are confident about is that risk management, while increasingly supported by sophisticated tools and techniques, is still a "people" operation. Those who make risk decisions often have dozens of years of experience and insight, something that is called "deep smarts" or organizational wisdom. And therein lies the challenge. In the United States, every eight seconds on average a member of the 76 million baby-boomer generation turns 65. So what does that mean? It means that a huge generation of employees has started to retire, taking with them years of knowledge and experience. Instead of viewing this as a negative, it can also be an opportunity to craft a workforce that is not committed to existing ways of doing business. It also offers the opportunity to practice something called *talent management*, arguably a term that is becoming trendy and overused.

Talent management is the process through which employers anticipate and meet their needs for human capital.[8] Without question, effective supply chain risk management benefits from having the right set of tools and techniques. Never forget, however, the importance of having the right people who know how to develop these tools as well as how to use them.

An ongoing challenge is finding employees who are capable of more than simply taking a tactical and limited view of the supply chain. This includes understanding how to incorporate risk topics with everyday business topics. Ideally, we want employees who not only understand the basics of supply chain management; they also need to have the ability to identify and anticipate potential risk outcomes. Over and over we see companies that have developed decentralized supply chain groups that are proficient at tactically managing transactions and material flow through a relatively uncoordinated network (something that itself presents risk). Understanding how various internal and external groups must collaborate to manage a value chain, including from a risk management perspective, is not as well understood.

What should we broadly look for in terms of human capital and risk management? First, employees who can view the supply chain holistically and in terms of linked processes are an absolute requirement. These employees must also understand their company's business model, including how risk management fits into that model. They must also understand how to manage critical supply chain relationships with a keen eye on risk management. A strong knowledge of cost and risk management techniques combined with the ability to engage in statistical analysis and fact-based decisions is also a necessity. The ability to walk on water would also be a definite plus. The knowledge and skill set required of tomorrow's supply chain professional is a bit more demanding than what we see today.

A challenge moving forward concerns how to acquire, assess, retain, and develop people with the knowledge and skills who are proficient at supply chain risk management. Table 3.1 provides some approaches for ensuring a continuous flow of highly qualified supply chain personnel. As this table shows, there are many ways to acquire and develop human resources.

As it pertains to how people view and perhaps even manage risk, interesting differences occur between genders. A body of research has concluded that women tend to focus on the odds of winning (the risk), while men focus on the payoff or reward. As a result men, on average, are more willing to take risks even when empirical evidence shows something to be inherently risky. Women tend to be more cautious in their actions and deliberate in their analysis. While at first glance this timidity may appear

TABLE 3.1

Talent Management Strategies for Acquiring and Retention

Approaches for Acquiring Talent	Approaches for Developing and Retaining Talent
Develop closer relationships with a select group of colleges to recruit interns and graduating students	Assign team leadership roles as a way to develop leadership talent
	Assess current knowledge and skill gaps by employee and position with customized development plans
Recruit management consultants who are leaving the consulting industry	Develop true mentoring programs
Recruit talent from other companies	Implement job rotational programs for high potential new hires
Recruit talent from the open market	Offer continuous and customized training programs to employees
Recruit talent from other internal functional groups	Develop career ladders that illustrate career paths
	Practice talent analytics, which involves adopting sophisticated methods for analyzing employee data
Recruit honorably discharged veterans	Create leadership development programs
	Reward personnel skill advancements
	Benchmark talent management best practices, including market salaries

to be a sign of weakness, it can also provide the correct reading of risk. And when women do select a course of action, they will work as diligently and often more successfully than men because they are selecting battles they are more likely to win.[9]

It is difficult to overstate the relationship between these four enabling areas and the successful pursuit of risk management strategies. The importance of people, organization, systems, and measurement will not diminish over time, making the need to build a strong foundation across these four areas a continuous challenge.

LINKING SUPPLY CHAIN RISK MANAGEMENT AND SUPPLY CHAIN STRATEGY

A second major part of building a risk management foundation is forging the critical linkages between risk management strategy and supply chain strategy development. At some point we expect to see risk management become an embedded consideration during the development of not only supply chain strategies but also corporate or business strategies.

The concept of strategy, at least on paper, is relatively straightforward. A strategy is a conceptualization of an organization's long-term objectives (which are aspirations) and purposes, broad constraints and policies that affect activities, and the current set of action plans (tactics) and near-term goals expected to help achieve an organization's objectives. An effective supply strategy consists of a number of parts—objectives, constraints, plans, and goals. A clear sign that an organization has achieved a level of maturity within a particular area is the presence of well-crafted strategies.

Making risk assessment a required part of any strategy development process makes sense for several reasons. Combining risk assessments with business strategies requires participants to consider risk issues. No longer can ownership for risk be deferred to another individual or group or an assumption made that risk will be considered at a later date. Making risk plans part of the strategy development process also helps embed risk management into the corporate culture. Participants begin to understand they cannot propose a strategy without a well-thought-out risk plan as part of that strategy. And these participants will come to realize they are the owners of the risk management process. Finally, making risk assessment part of any strategy development process usually does not require major changes to the strategy development process.

It should not be hard to grasp the logic behind identifying potential risks whenever strategies are developed. Many of these risks will qualify as constraints that have the potential to affect strategy success. The risk management approaches and techniques that are put in place should be viewed as tactics. The following identifies an important way to link risk management strategies and supply chain commodity strategies.

Integrating Risk Management with Commodity Strategy Development

Clear evidence of organizational maturity is the presence of well-thought-out strategies, which in supply chain management includes commodity or category strategies with suppliers. A purchase commodity or category is simply a like grouping of items or services. Some easy yet powerful ways are available to embed supply chain risk management directly into the commodity strategy development process.

A set of questions should be asked when crafting commodity strategies. The following provides a listing of the kinds of questions to consider when crafting a commodity strategy. It becomes relatively easy to incorporate risk management into this list. Performing a thorough strategy assessment, as demanded by the items below, is in itself a risk management technique. When crafting a commodity or category strategy, participants should ask

- What are the broad objectives of the strategy?
- What functional groups should be involved in strategy development?
- What are the internal customer requirements for this item or commodity?
- What are the current expenditures by commodity and supplier?
- What are the future volume requirements for this item or commodity?
- What is the performance of the current supplier?
- Who are the other potential suppliers in the marketplace and what are their capabilities?
- What other companies are competing for supply?
- Should we develop a domestic, regional, or worldwide contract?
- Should we rely on single or multiple suppliers?
- How should we divide the total spend if multiple suppliers are used?
- Do we want short-term or longer-term contracts?

- What are our pricing targets?
- What nonprice issues should we consider during contract negotiation?
- How do we measure supplier performance during contract performance?
- What specific services, such as inventory management or design support, will we seek from the supplier?
- What is the power relationship among the participants in the supply market?
- How can we promote continuous performance improvement?
- What type of supplier relationship should we pursue?
- Who will have responsibility for managing the supplier relationship?
- What specific contract clauses should we include?
- What is our negotiating strategy?
- What contract and supply chain performance risks do we need to consider?
- What risks may affect the supply market and suppliers?
- What risk prevention and mitigation actions should we put in place to address commodity and supplier risks?

Something that should increasingly be required is for commodity teams to include risk assessment plans as part of their formal commodity strategies. This forces commodity teams to assume responsibility for risk management rather than shifting that responsibility to another party. Recall from Chapter 1 that a risk assessment plan is an extension of a risk analysis. The risk plan is a document that defines known risks and includes descriptions, causes, probabilities or the likelihood of risk occurrences, costs, and proposed risk management responses. It can easily be included as part of a formal commodity strategy.

Commodity Risk Analysis and Plans. What does a risk analysis look like? The following provides some guidance regarding the format and content of a risk analysis report that could become part of a commodity strategy. These risk plans can also be developed independent of commodity strategies. To the best of our knowledge, no industry-accepted format exists regarding what a risk analysis should or should not include.

Section 1: This section includes an external intelligence report that describes in detail the supply market for the commodity/material. Who are the major suppliers and where are they located? Who are the major customers? What are the supply trends? Are there specific supply and demand price drivers? What is the overall competitive environment of the market for this commodity?

Section 2: This section identifies and categorizes risk(s), including a detailed description of each risk (i.e., not a generalization such as "potential supply disruption" or "bad weather").

Section 3: This section requires the development of a risk scenario map with each risk numbered and plotted on the map. The dimensions of the map can include the probability of a risk occurring and its expected impact if it were to occur.

Section 4: This section contains a comprehensive risk management plan that identifies risk management actions that describe how to manage the risks identified in Section 2. This section should also include a timeline that shows how and when to carry out risk management actions.

Section 5: This final section includes a listing of objective references and information sources about the demand and supply market for that item and supplier(s). It should identify why each information source is valuable. Particular emphasis should be given to information sources that are updated on a regular basis.

THE ULTIMATE RISK—IMPROVE OR ELSE!

A number of years ago executive management at a major steel producer presented an ultimatum to its subsidiary railroad units: Achieve a certain level of return on net assets (RONA) or risk divestiture from the parent company. Historically each railroad operated as its own business entity, and the parent company measured the financial performance of each unit separately. For the first time it became obvious that the different railroads needed to coordinate their activities across many different dimensions or face a very uncertain future.

This case illustrates how a strategic risk (divestiture from the parent company) was avoided through actions supported by stressing the four enablers described in this chapter—effective measurement, organizational design approaches supported by strong central leadership, capable human resources, and the extensive use of information technology. Combining these four enabling areas with strong central leadership changed the way this company conducted business with its suppliers.

Rallying around a Superordinate Measure

A key learning from this case involves the power of a seemingly ordinary measure (RONA) to become a superordinate measure that helps reduce risk at the corporate level (the measurement enabler). A superordinate measure is one that no single part of an organization can achieve on its own. By definition, the presence of superordinate measures demands that different groups work together or fail individually.

An emphasis on net asset return began when executive management challenged the railroads to develop new ways to meet or exceed an established RONA target. Figure 3.2 shows how this company measures RONA. A focus on this measure forced different functional groups to think about the big picture and how they each impacted that picture. The corporate decree to improve RONA forced functional groups to devise creative ways to increase earnings while simultaneously reducing assets and other liabilities.

Figure 3.2 also shows how various groups were responsible for different parts of the RONA equation. Accounting worked to improve accounts receivable and payable while marketing assumed responsibility for increasing revenue outside the parent company. Finance had responsibility for validating the savings and the numbers that populated the RONA equation.

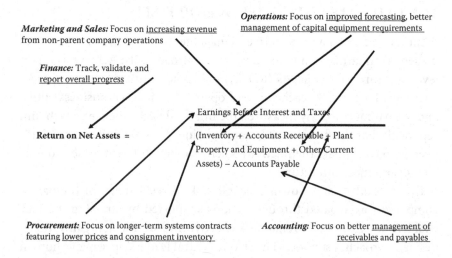

FIGURE 3.2
Managing return on net assets (RONA).

A small centralized purchasing group (the organizational design enabler) focused extensively on the denominator of the RONA equation. This group was responsible for developing innovative ways to manage spare inventory, which for the railroads represents a major financial commitment. This group radically changed how the railroads contracted with suppliers, changes that resulted directly in better financial performance and reduced risk.

Managers from the departments responsible for supporting the RONA target met regularly to share information and to discuss progress against their financial return targets. A sense of urgency surrounded these meetings simply because the risks were so high. The railroads viewed their asset return achievements as objective evidence of their contribution to the parent company.

Reducing Supply Risk through a New Approach to Contracting

The central purchasing group's approach to improving RONA involved the development of three-year systems contracts that featured consignment inventory. Previously, contracts with suppliers, if they could be called that, were developed by the individual railroads annually. Inventory consignment involved deferring payment for an item until a user at a railroad physically takes an item from a yard or warehouse and receives it into the railroad's inventory. The central procurement group developed 25 to 30 systems contracts, each covering around 25 items, with six suppliers. Contract renewal or renegotiation occurred every three years.

The procurement group redefined how the railroads sourced materials, a change that reduced financial and strategic risk. The previous procurement model featured each railroad issuing its own annual purchase orders. This model, which almost always featured annual price increases, was replaced by contracts that combined the volumes of the six railroads to realize more attractive prices and service. The central procurement group represented the interest of the six railroads not only when developing companywide contracts and but also during supplier negotiations.

While systems contracting is a radical departure for this company and its suppliers, it was not new to the head of the purchasing group. This individual had extensive experience with systems contracts and consignment inventory while working at General Electric (the human talent enabler). This highlights a major lesson here—transferring knowledge from one industry to another can provide tremendous benefits, particularly when dealing with a less-sophisticated industry.

Early in the process, purchasing announced that it would not accept higher prices due to any inventory carrying costs for the consigned inventory. During negotiation planning, purchasing used previous costs as a basis for price negotiation, although some suppliers attempted to add consignment costs into their purchase prices.

Why would suppliers agree to a systems contract that requires them to assume inventory carrying costs? The major incentive was three-year contracts that resulted in greater volumes that offset any consignment costs. Furthermore, the willingness of the railroads to take ownership of unused consigned material from suppliers at the end of each year reduced some supplier risk exposure.

Purchasing relied extensively on spreadsheet tools to analyze purchased items. Tying into the railroads' automated inventory databases, the purchasing group calculated actual usage and company-wide requirements, identified potential systems contract part candidates, and calculated inventory investment figures (the IT enabler). Spreadsheets were also used to analyze supplier-provided data before commencing formal negotiations. Computerized spreadsheets that retrieved data from various databases were invaluable when developing systems contracts. What became evident from this analysis was that a supplier may not be competitive across all items being considered on a systems contract, something that encouraged the development of "configured supply networks." Figure 3.3 explains the concept of a configured supply network with a simple set of data.

The purchasing group also conducted frequent meetings with the president of the railroads to report on progress. The president's agreement with the systems contracting approach sent a strong message to suppliers concerning the seriousness of this new way of doing business. Meeting with executive management also served to protect the president from being caught unaware or being unduly influenced when contacted by suppliers who preferred the old way of doing business.

Systems Contracting Benefits

Systems contracting featuring consignment inventory produced direct benefits for the railroads. First, purchasing achieved its primary goal of reducing inventory investment by almost 50%. Cash flow also improved

Assume suppliers provide the following quotes for supplying system contract items.
Step 1: Working row to row, identify the lowest price for each item.

Item/Part Number	Supplier A	Supplier B	Supplier C	Supplier D
442311 Gloves	*$1.26 (per unit)*	$1.65	$1.45	$1.29
338922 Wax	$7.45	$7.61	*$6.15*	$6.90
9963782 Glasses	*$2.10*	$2.54	$2.43	$2.91
746322 "D" Batteries	$.40	*$.30*	$.36	$.35
854471 Soap	$4.45	*$4.01*	$4.55	$4.50

Step 2: Calculate the total dollars based on annual volumes. Each cell = (price × volume)

Annual Volume	Supplier A	Supplier B	Supplier C	Supplier D
Gloves—80,000	*$100,800*	$132,000	$116,000	$103,200
Wax—5,250	$39,112	$39,952	*$32,287*	$36,225
Glasses—3,000	*$6,300*	$7,620	$7,290	$8,730
Batteries—30,000	$12,000	*$9,000*	$10,800	$10,500
Soap—8,000	$35,600	*$32,080*	$36,400	$36,000
Total	$193,812	$220,652	$202,777	$194,655

Step 3: Sum the shaded areas to arrive at a cost of $180,467, which is almost $13,000
less than the lowest cost Supplier A ($193,812).
Step 4: Configure the supply network to use different suppliers for specific items
to take advantage of price differentials.

FIGURE 3.3
Creating a configured supply network.

due to lower ordering and inventory carrying costs. The railroads also avoided or deferred price increases due to fixed pricing agreements. Systems contracts have also allowed for some downsizing as suppliers assumed responsibilities for delivering and placing physical inventory in storage at the rail yards and warehouses. In short, there was a great deal to like here.

While the parent company failed to achieve its own financial targets, the railroads achieved net asset returns of over 50%, prompting some to suggest that perhaps they should divest themselves of the parent company. This new way to operate, supported extensively by the four enablers outlined in this chapter, helped the railroads avoid their ultimate risk—their own demise!

CONCLUDING THOUGHTS

It is inevitable that the evolving discipline called supply chain risk management is going to be more complex, sophisticated, and demanding compared with what we face today. The development of risk management strategies must become more commonplace, and this demands building a foundation of people, systems, measures, and organizational design that supports risk management excellence. After thinking about the four enablers presented in this chapter, how many of us would logically conclude these enablers are not critical to effective risk management?

Summary of Key Chapter Points

- Perhaps one of the most underappreciated parts of corporate success, including successful risk management, is the role that organizational design plays.
- Supply chain risk management is rarely established as a distinct function within a company, although most organizations designate a "point person" with responsibility for risk management. Businesses have not yet agreed upon any typical way to integrate supply chain risk management into their decision-making processes.
- Various design features are ideal for including risk management as part of their scope, including sales and operations planning; collaborative planning, forecasting, and replenishment processes; executive responsibility; organizing around processes; and organizational work teams.
- Regardless of the type of information technology platform or software used, supply chain systems should have the ability to capture and share information across functional groups and organizational boundaries in real time or near real time.
- It is important to understand the important role that shared, real-time data can provide to risk managers. There is no question that "big data" is the next big thing in the IT world.
- Measurement supports making fact-based decisions about risk issues. Measurement is also an ideal way to communicate information, promote continuous improvement, convey what is important, and identify whether new initiatives are producing the desired results.
- A second major part of building a risk management foundation is forging the critical linkages between risk management strategy and supply chain strategy development.
- Making risk assessment plans a required part of any strategy development process forces participants to consider risk issues when developing strategies while helping to embed risk management into the corporate culture.

- A supply market risk analysis should include a supply market intelligence report, a categorization of identified risks, a risk scenario map, a comprehensive risk management plan that identifies specific risk management action, and a listing of references and information sources that include objective information about the demand and supply market.

ENDNOTES

1. Hamel, G., and C. K. Pralahad. "Competing for the Future." *Harvard Business School Press*, Cambridge, MA (1994), as referenced in Hellriegel D., J. W. Slocum, and R. W. Woodman, *Organizational Behavior*, South-Western College Publishing, Cincinnati (2001): 474.
2. Dumke, Daniel. Accessed from http://scrmblog.com/review/researchers-perspectives-on-supply-chain-risk-management.
3. Accessed from www.scm.ncsu.edu/public/cpfr/index.html.
4. Teach, Edward. "The Upside of ERM." *CFO*, November 2013: 44.
5. Lambert, Douglas M. "The Eight Essential Supply Chain Management Processes." *Supply Chain Management Review*, 8, 6 (September 2004): 18; Mackay, David, Umit Bititci, Catherine Maguire, and Aylin Ates. "Delivering Sustained Performance through a Structured Business Process Approach to Management." *Measuring Business Excellence*, 12, 4 (2008): 22.
6. Trent, Robert J. "The Use of Organizational Design Features in Purchasing and Supply Management." *Journal of Supply Chain Management*, 40, 3, (Summer 2004): 4.
7. Vance, Ashlee. "The Data Knows." *Business Week*, September 12–18, 2011: 70–74.
8. Cappelli, Peter. "Balance Your Talent Requirements." *Inside Supply Management*, 21, 10, (October/November 2010): 28.
9. Paskin, Janet. "Finding the 'I' in Team." *Bloomberg Business Week*, February 18–24, 2013: 78.

4

Strategic Risk

If there is one item that is the lifeblood of chocolate producers, it is cocoa. And if there is one item that presents a strategic risk to chocolate producers, it is cocoa. Around 70% of the world's cocoa crop is concentrated in five African countries, in a region that is not known as the most stable place to obtain raw materials. Furthermore, several years ago, a handful of traders took possession of almost all the cocoa beans in certified warehouses in Europe, raising legitimate concerns about commodity manipulation.

Even on a good day cocoa can be risky to grow as yields are lower compared with other crops, access to fertilizer is limited, and the cocoa crop is highly susceptible to pests. Not surprisingly, cocoa growers are increasingly shifting to more profitable crops such as rubber while younger farmers are reluctant to become cocoa farmers in the first place. None of this is good news to chocolate companies.

Major chocolate companies have come to the conclusion that they must work directly with farmers to introduce trees that increase crop yield, to eradicate pests and disease, to provide training and education, and to make sure farmers have access to fertilizer.[1] They are also working to make the financial model for growing cocoa beans more viable. These companies are trying to be proactive in the face of a strategic risk that can affect the success of a global industry.

This chapter addresses a specific category of risk called *strategic risk*. While countless corporate-level actions, decisions, and random events can create strategic risk, this chapter focuses on supply chain areas that have clear strategic linkages when not managed or anticipated properly. We will focus on strategic risk within three areas—strategic risk that results from new product failures, theft of intellectual property, and external intelligence failures. We also provide advice about how to minimize the strategic risk related to each area.

WHAT IS STRATEGIC RISK?

Before discussing various supply chain areas that can elevate risk to the strategic level, let's be clear about what we mean here. A good place to start is by understanding what is meant by *strategic*, a word that is one of the most overused terms in business. In fact, it is so overused that it often becomes difficult to know when something truly is strategic.

Something is strategic if it is necessary to or important in the initiation, conduct, or completion of a strategy or strategic plan.[2] Another perspective says that something is strategic if it relates to the identification of long-term or overall aims and interests of an organization and the means of achieving them.[3] However, all of the definitions tie into the notion that something is strategic if it has the ability or potential to affect the integrated whole, which means affecting an entire business or its continuity.

Now that we understand the word *strategic*, let's understand *strategic risk*. One perspective views strategic risks as those risks that are most consequential to an organization's ability to execute its strategy, achieve its business objectives, and build and protect value.[4] Another perspective views strategic risk as the current and prospective impact on earnings or capital arising from adverse business decisions, improper implementation of decisions, or a lack of responsiveness to industry changes or forces.[5] And a third view defines strategic risk as an array of external events and trends that can devastate a company's growth trajectory and shareholder value. This third view further categorizes strategic risks into seven major classes: industry, technology, brand, competitor, customer, project, and stagnation.[6] Consider the strategic risk to the producers of the lucrative *Fast and Furious* franchise. When Paul Walker, a major star of this series, was killed in an unrelated accident during the filming of *Fast and Furious 7*, Universal decided to indefinitely postpone the $200 million project, even though the movie was well into filming. The death of a film star during production is perhaps the ultimate strategic risk for a movie company.

Strategic risks are those that capture the attention of the board of directors. And these are the risks that make their way onto the 10-K report as enterprise risks. These risks have the ability to affect business continuity, erode a company's brand image, and adversely impact market share. As mentioned in Chapter 1, supply chain risks are increasingly becoming part of the enterprise risk listing, something that makes supply chain risk management a growing concern to executive management. The following presents three supply chain–related areas that minimize strategic risk when managed properly.

REDUCING STRATEGIC RISK THROUGH BETTER PRODUCT DEVELOPMENT

It is widely accepted that the successful development of products and services is an important part of what differentiates one firm from the next. In fact, a large body of literature has identified product development as a core process playing a major role that supports global innovation and competitiveness. The process of discovery, development, and commercialization of products and services is a major source for innovation and growth. It is also a process that when performed poorly has strategic risk implications, particularly as it relates to a company's reputation and brand equity.

This discussion is not about how to develop new products and processes. Other sources address that topic well. Instead, we look at how to make product development better from two perspectives. The first perspective presents a set of best practices during product development. The second looks at the emerging process of integrating new product development and risk management. Better product development means less business risk.

New Product Development Best Practices

A well-designed product development process can lead to many benefits, including reduced market risk, shorter development times, and first mover advantages that capture market share or create barriers to entry. An analysis conducted by *Industry Week* revealed hundreds of ways to develop products and services faster, better, and smarter. Our experience with leading companies suggests that the biggest improvements in product development are the result of a well-defined set of practices that requires a closer look.

Concurrency. Concurrency during product development is defined by two dimensions. The first is the simultaneous development of products along with the physical processes required to produce them. The second dimension involves the simultaneous rather than sequential involvement of functional groups during development. Sequential development (or what some refer to as linear development) features a "handing off" of work from one functional group to another, something that requires a time-consuming learning period after each hand-off. It also results in far too much work being handed back for revision when a later group finds a design to be unworkable.

What is it about a concurrent approach with cross-functional teams that is attractive? A concurrent approach requires cross-functional agreement throughout the development process, which minimizes time-consuming and costly design changes at later development stages. This approach also supports the interaction of competent professionals, something that usually leads to better decisions. Furthermore, concurrency offers opportunities for early customer and supplier involvement, accelerated learning as cross-functional team members learn simultaneously rather than sequentially, and the establishment of organizational rather than more limited functional goals. Some good reasons exist to pursue a concurrent approach to product and process development.

Early Involvement. Most executive leaders now appreciate the value of external involvement during product development, and they further recognize the need to involve suppliers and customers earlier rather than later in the process. Recently, a leading maker of appliances relied on a supplier to act as a system integrator for a complex module. The integrator assumed design leadership, selected the component suppliers, and managed those suppliers during development and production. For the first time this original equipment manufacturer (OEM) did not suffer product launch delays or cost overruns related to this module. Launch delays and cost overruns are key risks that are part of every product development project.

A meaningful relationship exists between supplier involvement on teams, including product development teams, and a variety of desirable outcomes. Teams that involve suppliers, formally or informally, are generally more satisfied with the exchange of information with suppliers compared with teams that do not include suppliers. These teams also note fewer problems coordinating external work activity with a higher reliance on suppliers to support a team's goals. Perhaps most importantly, external evaluators rate teams that involve suppliers as more effective with greater effort put forth toward their assignments or projects compared with teams where supplier involvement was lacking.

While early involvement with suppliers sounds easy, the process does bring with it some issues. Confidentiality of information continues to be a major concern when involving external organizations in something as strategically important as product development. Other concerns include not knowing how to pursue early involvement, maintaining too many suppliers for a given requirement, or external relationships that are adversarial rather than cooperative. Given the expected growth in product teams that includes suppliers and customers, overcoming any barriers to early involvement, each of which presents a clear risk to the success of the process, must become a priority. Fortunately, none of these barriers violates the laws of physics.

Use of Information Technology. Successful product development groups rely extensively on software to accelerate and improve the development process, another best-practice characteristic. Software is available that supports product and process development through design of experiments, quality function deployment (i.e., translating customer wants and requirements into design specifications), and the methodical assessment of design for manufacturability or assembly, something that is essential when taking a concurrent product and process development approach. And let's not forget about the importance of computer-aided design (CAD), computer-aided manufacturing (CAM), and rapid prototyping applications. Keeping with the software theme, failure mode effects analysis (FMEA) tools support the assessment of potential failures in a design or process, while TRIZ software supports a disciplined approach to complex problems that are encountered during development.

Without question we must include here additive manufacturing through 3-D printing. This rapidly evolving technology is expected to be a disruptive technology, especially during product development. Product designers at Ford, an early adopter of 3-D printing, are now creating prototype parts for testing in days rather than months. Besides saving time, which is the holy grail of product development, Ford is saving millions of dollars during development. On a new engine program, Ford will 3-D print most of the major components on the engine. According to a Ford executive, "When you learn about some of the ways additive manufacturing has positively impacted, and even in some cases saved product launches, you begin to understand its value. Our team is allowed so much more time to innovate and improve products because they have more opportunities to utilize multiple variations more efficiently."[7]

Linking R&D and New Product Development. The fourth element characterizing leading product development involves a direct linkage between R&D and product development. Far too often worthwhile innovations from the laboratory are not commercialized. Best-in-class companies have in place a process to develop and validate new technology and then link that technology to the product development process. While not all new technology emanating from R&D labs will be commercially viable, the linkage of one process to another ensures that market-ready technology can be designed into new products. When the pieces come together, it can create a difficult barrier for less-effective competitors to overcome.

Target Pricing. Many product development efforts will benefit from target pricing (sometimes called target costing). With a traditional approach to product development, we arrive at the selling price by combining product costs and adding a profit margin. Unfortunately, this approach often overstates the price a customer will accept or ignores what competitors are doing. It is an inwardly focused approach that does not consider the voice of the customer. Traditional pricing also tends to minimize the importance of cost management during product design.

Target pricing is a complete reversal of traditional pricing. Under a target approach, product development teams identify the price that customers are willing to pay for a product or service. After identifying a target price, profit margins are backed out to arrive at a product's allowable costs. If allowable costs are below current cost levels, then the design team must identify ways to remove or lower costs or to accept a lower profit margin. With target pricing, costs are something to manage rather than take for granted. Surveys of competitive pressures almost always conclude that the need to manage costs is relentless and severe. Why not manage costs early during product development?

Subtle Control. Years ago Taguchi and Nonaka conducted a study that identified "subtle control" as a powerful yet simple concept that executive leaders should routinely practice when teams develop new products.[8] Simply stated, subtle control is a delicate balancing act that seeks to ensure that teams and processes proceed as expected during development but without the need for blatant control or micromanagement by executive leaders. Effective subtle control, which begins to resemble an organizational art form, is a product development best practice.

Executives have many ways to practice subtle control. They can identify which projects to pursue, select leaders and members for development teams, create and manage the development process that teams will follow, require performance updates at regular milestones, and establish broad performance targets that teams use when establishing their individual goals. Subtle control recognizes that while empowerment is an attractive idea, relinquishing complete control of product development is not quite as attractive.

Design for X. A final practice mentioned here is something called *Design for X*. Industry leaders appreciate the power of the product development process to satisfy some important aspirations or objectives. The term X represents different aspirations the development teams consider even before beginning design work. Design for X aspirations can involve design for quality, reliability, serviceability, sustainability, end-of-life recycling, target price, assembly, cycle time, postponement, and even risk management. This is a powerful concept because it ensures that important objectives are considered early on during product development.

Integrated product development is a major source point for innovation and growth. We know that world-class companies endorse certain practices that differentiate their product development process from those that are not leaders. Fortunately, what makes up this set of practices is becoming increasingly clear. These practices will also go a long way toward reducing strategic risk exposure by improving the chances of successful product development. There is nothing quite like a product failure to tarnish a company's brand reputation.

Bringing New Product Development and Risk Management Together

A typical model of new product development projects, particularly at higher technology companies, features engineering teams working quickly and as fast as possible to develop a new product or technology. Then, at some later point, suppliers are selected to support the design. Oftentimes the selection of suppliers is not the most-thought-out part of the project, although few companies will actually admit this. When this is the case, it is up to the supply group after product launch to address supply chain risks as they materialize. This model, where risk mitigation takes precedence over risk prevention, is more common than we would like to admit.

It is challenging to locate case examples where companies explicitly address product development and supply chain risk simultaneously. While integrated product and process development is well understood, the integration of supply chain risk and new product development is not. The convergence of product development and risk management is an evolving practice that aligns well with the quality management principles of prevention and quality at the source.

What are the characteristics of a process that simultaneously brings together product development and risk management? First, supplier selection must happen early rather than later in the design process. From a risk perspective this means that selection teams will not only consider the traditional selection criteria they normally use when evaluating suppliers, but they will also consider risk factors.

Next, each cross-functional team involved in product development, with the help of the supply management group, will have responsibility for identifying a set of supply chain risks that may affect their part of the project. These risks are collected and categorized for easy access. Then, development team members will meet regularly to review not only product development progress but also discuss the actions taken to address these identified risks.

Taking this a step further, the teams will estimate the probability of each risk occurring and the impact that risk will have on product launch if it occurs. A common approach is to arrive at a risk rating that equates with high, medium, or low risk that reflects the combined probability and impact of a risk. Priority is then given to evaluating the higher risks to determine what action can be taken to reduce their probability and impact, thereby turning a higher risk into a lower designed risk prior to product launch.

A high-risk designation might result from relying on a critical material that is volatile in terms of price and availability. The design team might qualify a less volatile material that could serve as a substitute if necessary. Or, a design team might be uncomfortable with the use of a single supplier. Here, the supply group might prequalify a secondary source to mitigate that risk.

The objective here is to launch a product with as few unresolved risks as possible. Any outstanding higher risks at product launch are an admission by the development team that it has decided, for whatever reason, to accept that risk. Any accepted higher risks should be kept to an absolute minimum.

THE ART AND SCIENCE OF NOT GETTING CAUGHT BY SURPRISE

It would be quite a challenge to find an executive who says he or she enjoys getting caught off guard. Unfortunately, a lack of external intelligence can clearly lead to surprises that elevate strategic risk. External intelligence consists of the data, information, and knowledge that come from outside your company, including macroeconomic changes, legal and regulatory changes, industry trends and changes, supplier and competitor actions, social and labor force trends, customer expectations, technology innovations, and risk factors. Two challenges associated with external intelligence are that either we have information overload or we have our heads in the sand and fail to see what is going on around us. The underlying objective of external intelligence is to avoid surprises that mess up our lives.

A concept related to external intelligence is a company's EIQ, or external intelligence quotient. EIQ represents how well your firm collects, assesses, and acts upon relevant external data, information, and knowledge. This is a particularly relevant risk management concept. When a lower organizational level senses a significant event about to happen (or has happened), how quickly is it before that event is part of an executive discussion? Let's further subdivide external intelligence into two primary groups: market intelligence and supply market intelligence.

Consider a recent case with clear strategic risk implications. Next-generation smartphones are starting to feature biometric security devices to read fingerprints, which will enable a user to securely speed electronic payments and retrieve music, documents, and files from cloud storage. In 2012 Apple bought fingerprint-sensor company AuthenTec for $350 million, an amount of money that is pocket change to Apple.[9] Shortly after the purchase the newly acquired company stopped selling its product to HP, Dell, Lenovo, and Apple's nemesis, Samsung. If this technology proves to be a game changer, and many observers expect that biometrics will be a disruptive technology, Apple could secure a strategic market advantage over companies that suddenly saw their primary source of biometric technology disappear. External intelligence is intended to help avoid waking up one day and finding another company disrupting, perhaps irreparably, your supply market.

Market Intelligence. Market intelligence deals with everyday information about developments in the external marketing environment. It is very much oriented toward the customer and the downstream part of the supply chain. A market intelligence system is a set of procedures and sources used to obtain that intelligence. Most firms rely on market intelligence systems and marketing research to gain external knowledge about their markets. Companies can take a variety of steps to improve the quality of their market intelligence systems, including the following:

- Establish 1-800 phone number or web-based customer feedback tools
- Set up a customer advisory board made up of representative customers or the company's largest customers

- Purchase market intelligence information from outside providers
- Create a process to circulate relevant news stories and other information efficiently across your company
- Train and motivate the sales force to spot and report new developments in your marketplace
- Encourage distributors, retailers, and other intermediaries to pass along important market intelligence
- Buy a competitor's product
- Attend trade shows
- Subscribe to trade journals
- Review competitor brochures and websites
- Survey customers directly

Supply Market Intelligence. Supply market intelligence (SMI) is the output from obtaining and analyzing information relevant to a company's current and potential supply markets for the purpose of managing risk and supporting effective decisions. Four levels of supply market intelligence exist:

- **Macroenvironment.** This includes information about market dynamics and global economic conditions, world trade, demographics, political climate, environment, and technology.
- **Country level**. This category includes economic topics specific to a country, such as the location of free trade zones, the tax environment, labor availability, population size and trends, and regulatory bodies. It also includes information about cultural issues, political climate and stability, and national holidays.[10]
- **Industry and commodity.** This relates to the size and relative strength of industries and the worldwide users of commodities.
- **Supplier.** This category relates to the number of potential suppliers that exist, the products and services they provide, their size and capabilities, and location.

Let's provide several examples showing the linkage between a lack of supply market intelligence and strategic risk. Warnings from a supply market intelligence network could help avoid nasty surprises later.

Rare Earth Elements and Chinese Market Dominance. It is a safe bet that many companies rely on rare earth metals somewhere in their supply chain. And it is a safe bet that many supply chain managers do not grasp the strategic risk presented by rare earth metals. While these metals are found just about everywhere in the earth's crust, they are found in volume in only a limited number of commercially viable locations. Rare earth metals include 17 elements: 15 that are known as lanthanoids, plus scandium and yttrium. These elements are essential to defense products, such as missile guidance systems, as well as high-technology products. Unfortunately, at one point China controlled 97% of rare earth metal output around the globe. And the country has shown a willingness to play favorites, especially to its own companies. During a recent one-year period a rare earth composite price index increased 1,500%! Supply chain managers should immediately scan their supply chain, including subtier suppliers, to identify if and where these metals are consumed. Compounding this risk are the relatively few intermediaries that are capable of turning these elements into a semifinished form.

Guar and Fracking. The energy boom in the United States from hydraulic fracturing (fracking) has had an interesting secondary effect that involves, of all things, a bean. Guar, a beanlike vegetable grown mainly in India is used in countless products, such as ice cream, fertilizer, cattle feed, chocolate milk, hair conditioner, peanut butter, kitty litter, and cranberry sauce. It is what we call a ubiquitous item—it shows up just about everywhere. The real shock to the guar market occurred when oil and gas companies began using it as a key ingredient in their fracking solution. A single oil or gas well can require hundreds of acres of guar production. During the fracking process guar is mixed with water to thicken the fluid that is forced into the fractures of energy-bearing rock, thereby allowing oil and gas to seep out.[11] Needless to say, the guar market became quite volatile after the oil and gas industry became a major guar buyer. Many companies that use guar did not even know about the pending commodity market disruptions until it was too late to do much about them.

We could present dozens of cases that all lead to one conclusion—companies that are ignorant about their external environment elevate their risk exposure. Fortunately, thousands of resources are available that provide all kinds of valuable information and data. And unfortunately, thousands of resources are available that provide all kinds of external intelligence. What we are saying is there is almost too much information "out there." The need to converge on a set of trusted external sources of information has never been greater. Table 4.1 summarizes some carefully selected sources of external intelligence, including supply market intelligence.

TABLE 4.1

Examples of External Intelligence Sources

www.bls.gov	A repository of the Bureau of Labor Statistics of the U.S. Department of Labor, which is the principal Federal agency responsible for measuring labor market activity, working conditions, and price changes in the economy
www.census.gov	The Census Bureau serves is the leading source of data about the U.S. people and economy
www.finance.yahoo.com	An excellent source for current financial and company news, including company financial statements
www.google.com/alerts	A fast and easy way to set up alerts on any word, company, or topic
www.oanda.com/currency/converter	A comprehensive site for exchange rate and currency rate information. Parts of the site require a subscription fee
www.dnb.com	Dun & Bradstreet provides a variety of business, market, and supply intelligence products that tap into its vast database
www.thomasnet.com	A comprehensive website containing product, company, and industry news
www.industryweek.com	One of the best repositories of manufacturing information in the U.S.
www.cia.gov/library/publications/the-world-factbook/	The CIA World Factbook provides information on the history, people, government, economy, geography, communications, transportation, military, and transnational issues for 267 countries and entities
www.econsources.com	A portal to economic information on the web
www.politcalresources.net	Listings of political sites available on the Internet sorted by country, with links to political parties, organizations, governments, and the media
Unstats.un.org/unsd/demographic	This is the website for the United Nations statistics division
www.coface.com	A resource on country risk assessment, industry sector studies, and assessments of business climate around the world
www.investorsintelligence.com	A provider of research and technical analysis of stocks, currencies, commodities, and financial futures
http://www.havocscope.com/	A comprehensive site for information on global black market trade

PROTECTING INTELLECTUAL PROPERTY

International supply chains present a new set of risks compared with domestic supply chains. One area where this is of particular concern is the protection of intellectual property (IP). Intellectual property refers to creations of the mind, including inventions, literary and artistic works, symbols, names, images, and designs used in commerce.[12] Intellectual property falls into two main areas—industrial property and copyright. Industrial property includes inventions (patents), trademarks, industrial designs, and geographic indications of source. Copyright protects creative works by providing the copyright holder exclusive right to control reproduction or adaptation of such works for a certain time. A specific type of intellectual property theft involves counterfeiting, which Chapter 8 will explore.

Intellectual property protection is taken seriously, perhaps even for granted, in developed countries. The United States views intellectual property protection to be so important that the Constitution addresses this topic specifically in Article 1, Section 8, Clause 8. Intellectual property protection in the U.S. Constitution includes three kinds of intellectual property: patents, copyrights, and trade secrets. IP rights have since expanded to include trademarks. The Founding Fathers hardwired the importance of IP protection directly into the U.S. DNA.

Unfortunately, perspectives about the importance of protecting intellectual property differ widely worldwide. In fact, many social and legal topics are viewed differently from culture to culture, including what defines a normal workweek, child labor, environmental standards, treatment of women, and safe workplaces. Bribery, which some quaintly refer to as *facilitating payments*, and reciprocity, while both illegal in the United States, are often not illegal overseas.

While reliable figures are next to impossible to calculate, intellectual property theft costs U.S. companies anywhere from $80 billion to $250 billion annually, a figure that rises dramatically when calculating IP theft from a worldwide perspective. Some argue that these figures are even too conservative since IP theft, particularly when it includes counterfeiting, involves many hidden costs. What are the costs to individuals and society from fraudulent pharmaceutical products that harm patients? Precise figures are hard to come by since those that steal intellectual property are not too open about the scope of their activities.

Various studies have concluded that while intellectual property viola-tions occur in many countries, the Chinese are clear offenders. A report by AMR Research (now part of Gartner) involving 130 international compa-nies identified China as the "winner" in 9 of 15 risk categories, including intellectual property infringement, supplier and internal product quality failure, and security breaches. India earns top or second-place ranking in 9 of 12 problem areas. [13] Overall, earning the first or second spot in almost every problem area is a dubious distinction here. A major aerospace OEM is so concerned about intellectual property theft that its employees are not allowed to take electronic devices that store information into China when they visit on business trips. The company has also decided not to establish an R&D facility in the country, even though China is an emerging aero-space center. Chapter 12 will talk about macro country risk indicators.

The search for new markets and lower costs, particularly in emerging countries, forces supply chain managers to put more effort into protecting their firm's intellectual property. A fairly robust set of suggestions relates to protecting against intellectual property risk.

Develop Contracts. It is always a good idea to have a written and signed document that describes the expectations of the buyer and seller. This does not have to look like a U.S. contract. True international contracts exist if they adhere to the Convention on the International Sale of Goods (CISG), which the United States has signed. Be sure to perform a thorough reference check of prospective suppliers before entering a contractual agreement. And be sure the contract contains a well-written nondisclo-sure clause.

Limit Exposure to High-Risk Countries. Advanced, industrial coun-tries will have legal systems that can be trusted to treat foreign companies fairly, including the protection of intellectual property. On the other hand, developing countries may not offer effective legal protection against intel-lectual property violations. While we have a good idea about the worst offenders, we often disregard this information. If nothing else, knowing where the potential trouble spots are located can result in greater caution.

Register Trademarks and Patents. A piece of advice here is to transliter-ate your language into foreign characters to avoid embarrassing mistakes or usage by other parties. Transliterating means to represent or spell in the character of another alphabet. Although transliterating is not always easy, the process is critical to avoid infringement.

A second piece of advice is to move quickly to foreign trademark offices. In the United States, the first entity to use a trademark commercially owns the trademark. In China, the first to file the trademark owns that trademark. Speed is of the essence or else you might wake up to find someone else owning your trademark. Run, don't walk, to the trademark office. And be sure to retain a good trademark and patent lawyer.

A related concept to trademarks is something called *trade dress*. Trade dress, a form of intellectual property, is a legal term of art that generally refers to characteristics of the visual appearance of a product or its packaging that signifies the source of the product to consumers. Trade dress protection is intended to protect consumers from packaging or appearance of products that are designed to imitate other products, or to prevent a consumer from buying one product under the belief that it is another.[14] Registering a trade dress is a risk management approach for battling copycats and illegal knock-offs. According to one entrepreneur, "This [trade dress] is proving more effective than a patent because it falls under the Digital Millennium Copyright Act. There is no debate. Either they [copycat websites] have to remove it [my product] from the Internet and stop selling it or Google will shut them down."[15]

Divide Requirements. Another way to protect intellectual property is to disaggregate purchase requirements among several suppliers rather than providing a single supplier with access to an entire design. No single entity gets to see the big picture. Unfortunately, this approach clearly has the potential to increase costs, which begins to defeat the purpose of sourcing in emerging countries.

Establish Proprietary Assets in a Foreign Country. Some companies choose to establish operations in a foreign country or form a joint venture with a foreign partner, usually for the purpose of greater control. This may not be a realistic option for smaller companies or when sourcing raw materials or components. Rather, this is a more likely option when outsourcing finished products. Something that works against this option is reluctance by the finance types to take on new capital assets.

Seek Legal Support and Remedies. Legal experts can help identify preventive measures as well as legal remedies when intellectual property violations occur. These experts can also help develop contract language and nondisclosure agreements that address intellectual property protection, including the jurisdiction to resolve disputes. Legal groups can also help with the filing and protection of patents, trademarks, and copyrights outside a home country.

Some companies have taken an aggressive stand by pursuing IP violators wherever they exist. This strategy is both expensive and time consuming.

However, pursuing offenders can be an effective way to deter IP theft, particularly if a company has the support of a foreign country and an objective court system to back these efforts. It might also be good to earn a reputation as a company not to cross paths with.

Work with Trusted In-Country Representatives. A major risk internationally is not having a local representative that protects your interests. This representative, who should be intimately familiar with the region where business will occur, works in conjunction with a company's purchasing or sales office. Many smaller companies have found that using a trusted local representative is essential for protecting intellectual property.

Use International Purchasing Offices. Almost always staffed by a local staff, international purchasing offices can provide an invaluable in-country presence. These offices perform an array of tasks and services for the home office, including the evaluation of potential suppliers and an assessment of the supplier's willingness to respect intellectual property. The international purchasing office can also maintain a list of suppliers to avoid.

Move Production Back Home. Reshoring means bringing work back to a home country, usually due to economic factors, quality considerations, or the need to protect intellectual property. In the United States any reshoring that occurs is largely from China back to North America. Several U.S. companies have also closed facilities in Canada and Europe and shifted work back to the United States, although not because of intellectual property issues. For a variety of reasons, and beyond the scope of our discussion here, China is becoming less attractive as a sourcing location. A secondary benefit of reshoring, which occurs largely because of economic considerations, is to minimize the risk of intellectual property theft.

As this discussion shows, a variety of options are available that can prevent and mitigate risk due to intellectual property violations. Chapter 8 will expand on this topic by addressing fraud, corruption, counterfeiting, and theft.

WHEN STRATEGIC RISK BECOMES STRATEGIC REALITY

Supply chain risk managers and academics will be discussing the following case of strategic risk for many years. Anyone who is familiar with Class 8 (heavy-duty) trucks knows that Navistar has historically been a premier brand in that industry. A failed technological bet by Navistar's (former)

CEO has caused this once proud company to lose its luster. This case highlights the perils of strategic risk at its finest, or at its worst, depending on your perspective.[16] It is also an example of how a relatively small matter escalated into a near-death company experience. As if the technology failure was not enough, the company has also faced serious accounting issues with its auditor and the Securities and Exchange Commission.

To understand what went wrong, we have to step back in time. During 2000, the U.S. Environmental Protection Agency introduced sweeping new air quality regulations for diesel engines, requiring an eventual 90% reduction by 2010 in the amount of nitrogen oxide and soot these engines produced. New regulations even required a reformulation of diesel fuel to reduce its sulfur content. The total cost to industry of meeting these new regulations was estimated to be over $4 billion.

While these regulations may be daunting, this industry is no stranger to new regulations. The question became how to achieve these reductions? Truck makers had available to them at least three technological paths they could follow—two that were proven but still needed refinement and one that was in early stages of development. The proven technologies, which Navistar's competitors eventually adopted, included nitrogen oxide adsorbers (also called "traps") and a chemical treatment system called selective catalytic reduction, a system favored by European producers. Navistar's CEO, who had a reputation as a risk taker, decided to pursue an unproven technology. Navistar's engineers simply had to perfect the system, something that proved frustratingly elusive. And for whatever reason, the company neglected to develop a Plan B in the event the riskier path proved unworkable.

Navistar eventually committed $700 million, as well as its corporate reputation on exhaust gas recirculation (EGR), a technology that had limited application in engines. While other approaches eliminate nitrogen oxide by the use of a bulky chemical treatment system that requires an extra tank of fluid after treatment, EGR is radically different. With EGR the motor circulates exhaust gas back into the engine cylinders to burn the exhaust again, resulting in a cleaner, cheaper, lower-maintenance solution. Instead of simply meeting new environmental regulations, Navistar saw a way to differentiate its engines through the use of a radical new technology.

Navistar's competitors eventually outfitted their engines with a European-proven technology that relies on a catalytic converter and chemicals to break nitrogen oxide into nitrogen and water, an approach that has become the industry standard. The bottom line is that Navistar's engineers could not make EGR technology work, something that became painfully obvious around 2008. Even after Navistar's technical people informed the CEO they did not know how to make this technology work, he would tell them they had to come through and that he did not accept negative thinking. After a court eventually refused to allow Navistar to sell engines that failed to meet emission requirements, the company was forced to redesign its heavy trucks and engines while it still had the money to do so. In the interim, Navistar announced a deal to purchase engines and emission technology from Cummins that would enable its trucks to conform to EPA requirements.

This case illustrates how a company that spent decades building a world-class brand can lose brand equity and market share quickly. That, unfortunately, is the nature of strategic risk, which is why it is called *strategic* risk. A second lesson is that while thoughtful risk taking can be a good thing, risk taking without a backup plan is not recommended. When the stakes are high, make sure Plan B can be rolled out quickly. And finally, the corporate culture created by the CEO contributed to even greater risk exposure. According to one Navistar executive, the CEO "made it abundantly clear that anyone who disagreed with him was being negative to the point people figured it would be a career killer to point out problems in achieving it."

Navistar is paying dearly for this misplaced bet, although the company is slowly recovering under its new CEO. Financially the company has experienced declines in heavy-duty truck sales and higher costs due to engines that do not meet EPA standards.[17] What started as a way to gain competitive advantage almost destroyed an American icon. And how is the CEO doing that was behind this failure? He retired in August 2012 after receiving a $7.9 million payout, stock options, and other payments and benefits. Not a bad return for betting the farm—and losing.

CONCLUDING THOUGHTS

Without question strategic risk is the risk category that aligns historically with enterprise risk. To some it represents the ultimate risk category—one that has the potential to bring an entire company to its knees. Unfortunately, supply chain risks are increasingly becoming part of the domain known as strategic risk. And that is a level of recognition that should not be that welcome. When supply chain risks become strategic risks, the life of supply chain managers is only going to become more difficult.

Summary of Key Points

- Whatever definition of the word *strategic* is used, all tie into the notion that something is strategic if it has the ability or potential to affect the integrated whole, which means affecting an entire business or its continuity.
- Strategic risks are those risks that are most consequential to an organization's ability to execute its strategy, achieve its business objectives, and build and protect value.
- Supply chain risks are increasingly making the list of enterprise risks, something that makes supply chain risk management a growing concern to executive management.
- The biggest improvements in product development are the result of a well-defined set of practices including concurrency, early involvement, the use of information technology, linking R&D to product development, target pricing, subtle control, and Design for X.
- While integrated product and process development is well understood, the integration of supply chain risk management and new product development is not. The convergence of product development and risk management is an evolving practice that aligns well with the quality principles of prevention and quality at the source.
- External intelligence consists of the data, information, and knowledge that come from outside a company, including macroeconomic changes, legal and regulatory changes, industry trends and changes, supplier and competitor actions, social and labor force trends, customer expectations, technology innovations, and risk factors. A lack of external intelligence clearly increases strategic risk.
- A concept related to external intelligence is EIQ, or external intelligence quotient. EIQ represents how well a firm collects, assesses, and acts upon relevant external data, information, and knowledge.
- International supply chains present a new set of risks compared with domestic supply chains. One area where this is of particular concern is the protection of intellectual property. Intellectual property falls into two main areas—industrial property and copyright.
- The search for new markets and lower costs, particularly in emerging countries, forces supply chain managers to put more effort into protecting their firm's intellectual property.

ENDNOTES

1. Rai, Neena. "Chocolate Makers Fight for Farmers' Loyalty." *The Wall Street Journal*, May 30, 2013: B6.
2. Accessed from www.merriam-webster.com.
3. Accessed from http://oxforddictionaries.com/us/definition/american_english/strategic.
4. Frigo, Mark, and Richard Anderson. "Strategic Risk Assessment: A First Step for Improving Risk Management and Governance." *Strategic Finance* (December 2005): 25–33.
5. Accessed from http://www.in.gov/dfi/RiskMatrixDef.pdf.
6. Slywotzky, A. J., and J. Drzik. Countering the Biggest Risk of All. *Harvard Business Review* (April 2005): 78–88.
7. Hessman, Travis. "Mastering the Hybrid Factory." *Industry Week*, June 2013: 12–15.
8. Takeuchi, Hirotaka, and Ikujiro Nonaka. "The New New Product Development Game." *Harvard Business Review*, 64, 1, (Jan/Feb 1986): 137.
9. Mawad, Marie, and Adam Ewing. "Apple Sets Off a Biometrics Arms Race." *Business Week* (August 22, 2013): 41–42.
10. Handfield, Robert. *Supply Market Intelligence*. Boca Raton, FL: Auerbach Publications (2006): 175.
11. Dezember, Ryan. "Farmer Says: Hitch Your Wagons to Some Guar." *The Wall Street Journal* (November 25, 2011): C1.
12. Accessed from http://www.wipo.int.
13. Anonymous. "China Contributes the Most Risk to International Companies." *Inside Supply Management* 20, 2, (February 2009): 8.
14. Accessed from www.wikipedia.com.
15. Accessed from http://www.nfib.com/business-resources/business-resources-item?cmsid=62517.
16. Public domain information, including Muller, Joann. "Death by Hubris? The Catastrophic Decision That Could Bankrupt a Great American Manufacturer." *Forbes*, August 20, 2012. http://www.forbes.com/sites/joannmuller/2012/08/02/death-by-hubris-the-catastrophic-decision-that-could-bankrupt-a-great-american-manufacturer/.
17. "Navistar International Second-Period Loss Swells on Decline in Truck Sales." *The Wall Street Journal* (June 11, 2013): B4.

5

Hazard Risk

In July 2011, a tsunami off the coast of Japan caused human tragedy on an almost unimaginable scale, killing thousands and rendering almost a million people homeless. Also in July, flooding in Thailand forced the closure of 9,800 factories and left more than 660,000 people unemployed. An untold number of people around the globe were impacted by this natural disaster, both personally and economically. In the automotive industry, approximately 6,600 autos were not built per day. In the camera industry, Nikon's net loss of sales equaled $786 million, while Canon experienced over $600 million in losses. With over 45% of all computer hard drives produced in the affected area, many computer manufacturers had to delay product launches, which resulted in lost sales and a doubling of hard drive prices.[1] Unfortunately, examples of hazard risks are not hard to come by.

Hazard risk pertains to random disruptions, some of which involve acts of God. Additionally, hazard risk has traditionally involved liability torts, property damage, and natural disasters. In this chapter, we will address how corporations mitigate hazard risk, whether it's a natural disaster, fire, malicious behavior, product tampering, theft, or acts of terrorism. The chapter will cover definitions, options, and variants of traditional and emerging global disruption insurance products. Later chapters will present other tools and approaches that support the effective management of hazard risk.

THE TRADITIONAL WORLD OF HAZARD RISK AND INSURANCE

With global supply chains expanding to support growth initiatives, many companies are entering into indemnity and other types of partner

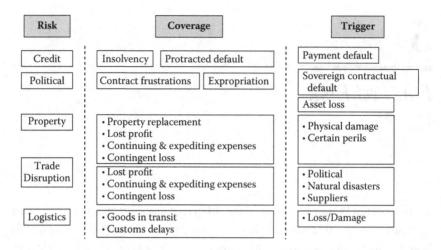

FIGURE 5.1
Risk transfer tools—the business of hazard insurance. (Source: Eric Wieczorek, former risk director-navigant, "Supply Chain Risk Management" presentation at MAPI, Manufacturing Alliance for Productivity & Innovation's Council Meeting, 2013.)

agreements in an effort to mitigate risk through risk pooling, sharing risk, and disruption insurance. Insurance may not be the total solution to mitigating risk within this category, but many new and improved business and trade disruption insurance products and services are emerging every day.

Before we move into definitions, options, and variants of risk transfer packages, we'll begin with a glimpse into the landscape of this traditional risk mitigation arena. We'll use Figure 5.1 as a reference point for our entire chapter discussion. This figure includes three columns. From left to right are the risks, the types of coverage, and finally the triggers or drivers behind traditional hazard event mitigation insurance.

The first risk in Figure 5.1 is credit risk. The trigger event is normally payment default, and the coverage is normally geared toward mitigating for insolvency or protracted default by a partner or customer. Next are political risks. The trigger events here are normally sovereign debt default or asset loss and coverage that revolves around contract frustrations and expropriation. Within the insurance coverage area for political risks, contract frustration tends to revolve around corporate loans, the lenders' interest, or project financing of equity and debt. Also within this coverage area, the bulk of expropriation revolves around fixed investments (equity) and possibly mobile assets, such as stocks.

Continuing down the left side of Figure 5.1 is property risk. The triggers in this category are physical damage or loss and additional perils or hazards that are catastrophic. Coverage works to mitigate property replacement, a loss of profit, expedited and continuing expenses, and a contingent loss. Next is trade disruption. How is trade disruption different from property or business interruption? It's a bit broader in nature, such as the Thai floods that may prevent the delivery of key equipment or supplies. Thus, with trade disruption solutions in place, customers can attempt to recover lost profits, and potentially expedited and continuing recovery expenses and contingent expenses as well. And finally, we have logistic risks. The trigger events are straightforward, such as physical loss or damage while in transit. Coverage works for goods in transit anywhere in the world and especially for customs delays beyond a company's control.

As our discussion moves forward within this traditional risk industry, it is appropriate to provide a few additional terms and definitions:[2]

- Hazard—a game of chance; a source of danger; a chance event or accident
- Liability—the quality or state of being liable; pecuniary obligations
- Liable—obligated according to law or equity; exposed or subject to some adverse contingency or action
- Tort—a wrongful act for which a civil action will lie, except one involving a breach of contract

The following summarizes various kinds of supply chain insurance coverage.

First-Party Commercial Property Insurance

This is by far the most common form of coverage, which includes business interruption coverage. It typically reimburses the insured for profits that it would have earned had it not suffered a business interruption. The coverage normally pays for lost profits resulting from some type of interruption in the supply chain. The insured does not necessarily have to suffer an actual loss within its enterprise or supply chain. This was evident during the Thai floods. Many manufacturing buildings were not inundated during the flood. However, the entrance and exit for almost all the facilities built in the 100-mile flood plain were not accessible, and only a small amount of material got into and out of the facilities, subsequently affecting downstream contractors and original equipment manufacturers. Be aware that some policies only cover losses due to disruptions at the facility of a "direct" supplier or customer. This implies only first-tier suppliers or direct customers. Coverage within this solution should be thoroughly evaluated by the insured and perhaps their insurer on a regular basis.

Cargo Insurance

This coverage is common considering the vast amount of goods that are transported. It covers damage to or theft of the goods while in transit via air, ground, or water. For one example, containerization has grown exponentially due to global trade growth across all industries as well as the development of megaships that can move thousands of containers at a time. One of the most frequent claims involving containers involves theft, with an estimated $30 billion in losses being reported annually over the last several years. This coverage protects goods on ships, on trains, at airports, and virtually anywhere goods are in transit.

Cyber Insurance

Cybercrime is increasing in frequency and is a growing concern for supply chain managers, CIOs, and CFOs around the globe. Why? Because when a company's IT system is down, the company is not taking orders, making products, shipping products, or billing customers. In short, the company is not functioning. That constitutes a disruption. And let's not forget the seemingly endless stories of cybertheft.

This form of coverage is still in its infancy but has been used already with high-profile incidents with retail and consumer goods companies. The key to reviewing this type of risk coverage is to not only consider the impact on the insured organization, but also the impact on suppliers, customers, and their customers. These insurance packages will grow in stature and complexity for many years to come.

Business Interruption Insurance

According to the American Insurance Association, business interruption coverage is part of broader commercial insurance policies.[3] This coverage is most commonly found in commercial property insurance policies and is triggered when one of more of the following conditions are met: First, physical damage to the premises is of such magnitude that the business must suspend its operations. Second, the physical damage to property caused by a loss is covered under a company's insurance policy, and that damage totally or partially prevents customers or employees from gaining access to the business. And third, the government shuts down an area due to property damage caused by a peril covered by the company's insurance policy that prevents customers or employees from gaining access to the premises.

Something to keep in mind is that business interruption coverage is limited. After a mandatory waiting period, coverage provides for lost net income, relocation expenses to a temporary site, and ongoing expenses such as payroll. Coverage is available only for as long as it is necessary to get a business running again, usually not longer than 12 months.

Contingent Business Interruption Insurance

According to Zurich Insurance, contingent business interruption (CBI) insurance "reimburses a company for lost profits and other possible transferred risks, such as necessary continuing expenses, due to an insurable loss suffered by one or more of its suppliers or customers."[4] Recall that traditional business interruption insurance compensates policyholders for losses resulting from damage to their own property. Contingent business interruption insurance allows a policyholder to transfer certain risk losses to a third party. This type of insurance is also commonly referred to as dependent properties coverage. Similar to business interruption insurance, CBI is triggered by certain events such as a direct physical loss or damage to a dependent property (supplier or customer), the loss or damage is caused by a covered cause of loss, and the loss results in a suspension of operations at a covered location.

Trade Disruption Insurance

Trade disruption insurance protects against lost profits caused by disruptions in the supply chain where there is no physical loss or damage to the insured's or its suppliers' assets. It normally covers property, marine, and political risk. Again, physical loss is not required and the loss does not need to be location specific. It normally covers a loss for up to 12 months resulting from termination of a project or nondelivery of goods, in areas such as loss of profits, extra costs, and expenses. This product normally fills gaps in coverage left open by other solutions such as business interruption or cargo insurance. It also tends to cover losses from political and economic exposures such as confiscation, expropriation, or nationalization of the insured's assets.

Also included in trade disruption insurance is a law/order decree result-ing in something called *selective discrimination*—forced abandonment of a provider; forced divestiture; and cancelation of licenses, operating permits, or concessions. And finally, it tends to cover political violence, including war, insurrection, strikes, riots, terrorism, and civil commotion. If the insured purchases or produces product overseas and if payments are made locally in that country, coverage can include currency inconvertibil-ity (currency conversion and exchange transfer) that can protect against repatriation of profits, dividends, and other remittances.

This product has been well received. It provides many benefits in terms of risk mitigation, including immediate protection for emerging markets with a global scope of coverage.[5]

Global Logistics Insurance

A final coverage we'd like to address is global logistics insurance, a deriva-tion of the cargo insurance profiled earlier. Because most supply chains are now global, coverage packages have expanded to insure additional threats or points of peril due to worldwide operations. Just as contingent business interruption insurance has expanded beyond the original scope of regular business interruption insurance, so too has insurance expanded to include additional elements and activities within a worldwide supply chain. A sample of the additional kinds of global logistic insurance that are available includes the following:

- Importers and shippers can purchase insurance for cargo and for cus-toms bonds. A customs bond is a type of insurance that someone who is importing goods must have, in which they promise to pay any nec-essary taxes and obey the rules for importing goods.[6]
- Factories can acquire property/casualty insurance along with bailee liability protection. A bailee is a person or party to whom goods are delivered for a purpose, such as custody or repair, without transfer of ownership. A bailee possesses property that belongs to someone else.
- Freight forwarders and shipper's interests can be covered along with the availability of cargo, bailee liability protection, and professional liability protection.
- Consolidators can purchase cargo liability for bills of lading and warehouse receipts.

- Foreign ports and terminal operators can secure coverage for workers' compensation, terminal operator liability, property coverage, and stevedore liability (a stevedore is a person employed or a contractor engaged at a dock to load and unload cargo from ships).
- Ocean carriers can obtain hull insurance.
- Customs brokers can obtain customs bonds.
- U.S. ports and terminals can obtain coverage for stevedore liability, operator liability, and property protection.
- Trucking companies and rail lines can add motor truck liability, shipper's cargo insurance, and bailee liability protection.

Before we provide additional context into how insurers evaluate and quantify the coverage we discussed earlier, we felt it would be beneficial to explore the level of underinsured nations exposed to disasters. The Center for Economic and Business Research (CEBR) and Lloyd's of London collaborated on a research project to identify insurance coverage associated with countries rather than companies.[7] The research methodology focused on the amount of money spent on insurance premiums, which was then subtracted from the expected loss from natural disasters. That outcome was then adjusted to account for differences in gross domestic product (GDP) levels and then benchmarked as a percentage of the country's GDP.

This research revealed that 17 countries were "significantly underinsured" and that these countries had a combined average annual insurance shortfall of more than $168 billion. Of the 17 countries on Lloyd's underinsured list, 8 are in Asia. China alone is underinsured by more than $79 billion, making it the largest underinsured country in the study. Table 5.1 profiles the outcome of the research in terms of the worst underinsured countries to the most insured.

This research revealed additional troubling issues as it pertains to reviewing natural disasters and their impacts on both businesses and the public. The death toll from the 2004 Indian Ocean tsunami was 220,000 people, but the cost to the insurance industry was only $1 billion. A Lloyd's of London insurance chief executive stated that he hoped this research would stimulate a debate on how governments, businesses, and the insurance industry manage the risks of a natural catastrophe, particularly on the merits of risk transfer versus the use of public funds to cover costs. Lloyd's made an explicit comparison between how different countries fund disaster recovery and concluded that countries have drastically

TABLE 5.1

Underinsured Countries

Country	Insurance Coverage	Country	Insurance Coverage
Bangladesh	−2.6	Russia	0.3
Indonesia	−1.7	Japan	0.4
Vietnam	−1.4	Israel	0.4
Philippines	−1.4	France	0.4
Egypt	−1.4	Sweden	0.4
Turkey	−1.3	Argentina	0.4
India	−1.2	Italy	0.6
Nigeria	−1.1	Ireland	0.8
China	−1.1	South Africa	1.0
Chile	−1.0	Taiwan	1.0
Saudi Arabia	−0.9	Spain	1.1
Mexico	−0.7	Denmark	1.4
Brazil	−0.5	Australia	1.4
Thailand	−0.4	UK	1.6
Colombia	−0.2	Austria	1.7
Poland	−0.2	Germany	2.1
Hong Kong	0.0	USA	2.5
Singapore	0.1	Canada	2.5
UAE	0.1	South Korea	2.6
Malaysia	0.2	New Zealand	3.1
Norway	0.3	Netherlands	8.0

Source: *RIMS Magazine,* "Underinsured Nations Exposed to Disasters," Emily Holbrook, March 2013.

Note: These are relative rankings: Negative values mean significantly underinsured as a country; positive values mean better insured.

different approaches to disaster readiness. The greatest difference is the amount of money paid by taxpayers toward each country's recovery from catastrophic events versus funds paid by insurance carriers.

QUANTIFYING TRADITIONAL HAZARD RISK INSURANCE REQUIREMENTS

As we mentioned in Chapter 1, risk appetite reflects the degree of risk that an organization or individual is willing to accept or take in pursuit of its objectives. This can be measured in terms of both quantitative and qualitative dimensions. Some also refer to this concept as *risk tolerance* or *risk propensity*, a topic that is well grounded in the financial community. Finance experts view risk appetite as reflecting the type of risk that an institution or individual is willing to undertake in pursuit of a desired financial performance. Calculating a risk appetite in financial terms is what the insurance industry calls a *risk quantification engagement*. This engagement is a key element of the insurance industry's approach to deciding whether a client has too much or not enough insurance based on an assessment of that company's risk tolerance, risk policies, and programs.

This section of the chapter discusses what a typical quantification engagement looks and feels like and the outcome from those types of assessment projects. The integrated insurance cycle includes an "as-is" or risk quantification, leading to a potential new set of insurance solutions, possibly through stress testing, then an execution phase, and finally an operational mapping and risk identification phase. For purposes of illustration, we'll concentrate on the first phase, which is risk quantification.

The backdrop to this engagement normally begins with an internal dialogue, sometimes driven by the chief risk officer or the CFO's office, revolving around the question of enterprise risk exposure and the review of total insurable values. This dialogue pits the total insurable values against the amount of interruption and other insurance coverage the company has in place. The dialogue is basically driven by whatever discipline or department is responsible for enterprise risk management.

The initial in-house meeting usually moves quickly into a set of questions of discovery such as, Is there a difference between our reporting values to our underwriters and our reporting exposures to management? The dialogue should cover additional points such as, What is our present premium expense? What are our present policy limits and sublimits? Have all our exposures been identified? These exposures are generally discussed and can address areas such as a company's global markets, divisions, products and services, facilities within the supply chain, the supplier base, IT systems, and intellectual property. The discussion then migrates into additional executive questions such as, What does our business continuity plan (BCP) look like? Have we exercised our BCP, and if we have, how did it play out? Do we need to update our BCP? When these questions have been answered, the CFO's office or a divisional controller might make the injection that an existing insurer or another third party should be brought in to provide a risk quantification analysis. When that decision is made, the as-is portion of a risk quantification engagement commences.

The complexion of a typical risk quantification engagement often involves a kickoff meeting with executives from several disciplines or divisions. The meeting objectives and agenda are not at all trivial. In fact, this process can be quite complex. This process, starting at the kickoff meeting but extending into subsequent meetings and assignments, will involve a review of existing insurance coverage across an enterprise or division, a review of product sales and margins through forensic accounting of profit-and-loss statements, a discussion of new products and services and forward marketing strategies for the next several years, and the identification of products or portfolios with the highest contribution and risk.

This part of the quantification process should include *what-if* scenarios to stress test a company's operational processes. Frequent meetings occur to report on evaluations and recommendations. Risk quantification engagements can take anywhere from 8 to 15 weeks, based on the scope and scale. A divisional analysis will invariably be shorter than an enterprise-wide analysis.

The focus again is normally to first establish a maximum foreseeable loss (MFL) for the division or enterprise. The MFL equals the worst scenario from the what-if scenarios. The typical MFL includes five key components: the products' contribution margin, existing mitigation plans, finished or work-in-progress inventory, tax impacts, and expected continuing expenses for the duration of a loss.

Upon agreement on and analysis of the MFL definitions, the next step in the engagement is normally the detailed examination of a plant, a division with all its facilities, or the entire enterprise. For illustration purposes we'll discuss an engagement involving a division with several facilities. We'll use Figure 5.2 as a reference to describe the basic elements of a risk quantification engagement.

The basic steps are exercised for every fixed asset in the project, which Figure 5.2 illustrates. Let's take a more detailed walk through this figure:

1. Calculation of value-at-risk (VAR) first takes place. This is normally the annual profits for the asset or division under scrutiny.
2. A primary backup plan or site is usually identified if the asset is a plant or a contractor. This backup plan tends to reduce the overall VAR.

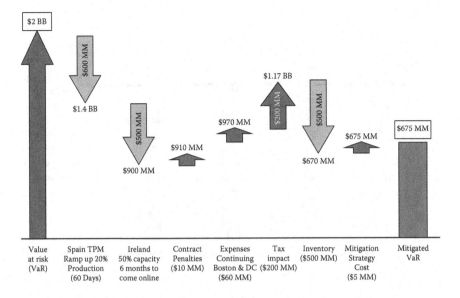

FIGURE 5.2
The risk quantification engagement. (Source: Eric Wieczorek, former risk director-navigant, "Supply Chain Risk Management" presentation at MAPI, Manufacturing Alliance for Productivity & Innovation's Council Meeting, 2013.)

3. Any additional backup facilities identified will further reduce the VAR.

4. Any contract penalties for service delays for existing and future customer contracts are identified and will potentially increase the VAR.

5. Any continuing expenses during the recovery period are identified, codified, and assigned to the overall VAR.

6. Any tax impacts are calculated based on the time-to-recovery (T-t-R) metric identified in the business continuity plan or scenario plans, which will also increase the VAR.

7. Any inventory in the pipeline and supply chain will reduce and mitigate the VAR during the T-t-R period.

8. Any new mitigation strategy costs to be incurred will increase the overall VAR.

9. Finally, on the far right of Figure 5.2 is the final mitigated VAR for that particular asset or product or division.

As the quantification project plays out, the ultimate goal is to identify what the maximum foreseeable loss might be, compare that to the risk appetite or tolerance of the organization based on their present level of

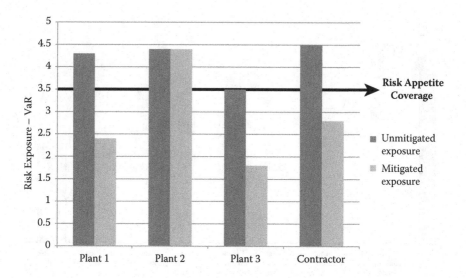

FIGURE 5.3

Risk quantification output. (Source: Eric Wieczorek, former risk director-navigant, "Supply Chain Risk Management" presentation at MAPI, Manufacturing Alliance for Productivity & Innovation's Council Meeting, 2013.)

insurance coverage, and decide if it's prudent to increase coverage or reduce coverage.

The output of this engagement is illustrated in Figure 5.3. The risk levels are articulated on the left or Y axis, low to high, and the nodes or assets within the scope of the project are identified on the X axis. The units of measure normally used in these engagements are either annually demonstrated profits or planned annual budgeted profits going forward. There are normally two pillars for each node or asset. The first pillar, or the darker shaded one, is the unmitigated exposure or VAR. Unmitigated risk, in insurance terms, generally means that the profit plan for this particular asset has no risk response plan, as the insurance investigators see it, in the event of a disruption and is therefore at risk. The second pillar, the lighter shaded one, is the mitigated exposure or VAR. Again, in insurance terms, this column represents the mitigated exposure for this asset, which normally means a risk response plan has been developed to ensure that profits continue in the event of a disruption and is supported by some type of insurance coverage. Both columns profile the profit exposure and neither eliminate risk. However, the lighter shaded, mitigated columns depict a reduced exposure to risk for the organization. The line passing horizontally across the entire graph is the present property, tort liability, continuity business interruption, and other insurance coverage the company maintains at the time of the quantification risk engagement.

The final, difficult portion of these engagements is to discuss what products, services, plants, or divisions have an exposure above the coverage line or risk appetite line. The insurance industry regularly uses the term *risk appetite* to denote the amount of insurance coverage a company has paid for to recoup losses or damages, based on the risks the company sees and how aggressive they want to be in terms of taking risks and supporting those risks with mitigation plans and insurance. As seen in Figure 5.3, when exposure or VAR lies above the coverage of risk appetite line, that situation drives the dialogue between the organization and its insurer or consultancy in an attempt to decide on the next steps. Those next steps normally include spending time and funds to develop more robust mitigation policies, procedures, and programs or increasing coverage to mitigate the exposure and risk. A company can also elect to assume some risk through self-insurance. And finally, scanning Figure 5.3 and viewing the differences in column heights, we'll attempt to provide a brief explanation in the context of these quantification tools and techniques.

The difference in column heights, which is profit exposure or VAR, per asset, is the difference between VAR exposure identified without or with risk mitigation strategies in place by the company. For instance, for Plant 1, there is obviously a set of mitigation plans in place to be used in the event that the plant is disrupted, because the lighter shaded column (mitigated) has much less VAR or exposure than the dark or unmitigated column. When the insurance experts complete the engagement, they calculate the total company VAR or exposure versus the amount of insurance coverage the company maintains at the time of the engagement and use the horizontal line to "make a point" about what assets have more VAR or exposure than the amount of insurance coverage. The bottom line of a risk quantification engagement is to discern whether a company has too much or not enough insurance based on the risks identified, mitigation plans in place and the company's risk appetite, which is indicative of their existing coverage level.

Outcomes from this type of financial risk quantification engagements and the benefits tend to be as follows:

- The calculation of maximum foreseeable losses
- A root cause analysis that highlights risk exposure

- A definitive picture of risks and a compelling reason to act to either mitigate those risks, temporarily or permanently, by buying more or less insurance, better prepare for disruptions or other risks, or continue to investigate how to mitigate the risks
- Assistance in obtaining capital funding for building redundancy and resiliency within the supply chain
- A positive restructuring of the insurance portfolio

LOOKING AT THE THAI FLOODS THROUGH A RISK QUANTIFICATION PRISM

Let's finish this chapter by revisiting the floods in Thailand that were mentioned at the start of the chapter, looking through the prism of risk quantification of what insurers might have recommended if they were asked to review the supplier base by the original equipment manufacturers (OEMs). First, let's focus on the event itself. The disaster followed with precision the four stages of a disaster, which are denial, severity, blame, and resolution.

During the *denial* phase, several days after the event, the media were told that minor disruptions would occur due to logistical issues. Aside from that, business would continue thanks to inventory in the supply chain. Next, about a month later the media began to hear and report on the *severity* of the situation. It became evident that industries would see supply shortages cause by the destruction of the work-in-progress inventory at the time of the event.

In the next phase, the media heard about *blame* as industries faced further shortages due to reduced production capacity as a result of production assets destroyed, factories being unable to restart due to power supply limitations, or a lack of raw materials from their own suppliers. Finally, industry members were saying after six months from the initial event that we would see further constraints while destroyed assets and factories got repaired or replaced and new capacity came online. This is the *resolution* phase.

What was going on during this event behind the scenes? Firms directly hit by the supply disruption were frantically canceling orders and attempting to book orders from alternate suppliers. Unfortunately, manufacturing capacity was already scarce before the event. And, qualifying and testing parts from new suppliers is time and resource intensive. Many alternate suppliers identified after the flood were, unfortunately, also located in the same flood plain region of Thailand (Chapter 12 will discuss the dark side of these industrial clusters). Furthermore, manufacturing yield and quality with new suppliers proved not to be the same as with qualified, incumbent suppliers. The inevitable outcome from this scarcity of supply was a dramatic increase in prices across a range of industries.

With the benefit of hindsight, what could insurers or consultancy organizations have done if they were asked to exercise a risk quantification engagement? Prior to the event, insurers or risk consultants would have exercised a risk identification, an as-is analysis, and a quantification engagement. They would have performed a thorough evaluation of the supply base and would have seen the exposure in terms of capacity, asset liability, and especially the geography and location of the entire industry supplier base.

As the event unfolded, the OEMs, at the insistence of their risk experts, could have rapidly activated a previously tested business continuity plan, which would have included secondary suppliers identified through the risk quantification exercise. Presumably, these suppliers would be located outside the impact zone. These companies would have also leveraged certain articles inside their newly purchased contingent business interruption and trade disruption insurance coverage packages that emanated from the risk quantification exercise.

After the event the OEMs would also have performed a post-event analysis and improved their initial risk plans through lessons learned. These companies might also have bolstered their mitigation plans and insurance coverage as well.

CONCLUDING THOUGHTS

Traditional hazard risk has been with us since the dawn of time and will be present until the end of time. One of the best ways, but not the only way, to mitigate these risks has been insurance coverage. Many existing and emerging packages are becoming available every day. Between the coverage instruments and the sophistication of insurers' risk quantification capabilities, it might be wise to consider these mitigation packages and processes on a regular basis as supply chains become more uncertain and complex.

Summary of Key Points

- Traditional hazard risk pertains to random disruptions, liability torts, property damage, and natural disasters that are outside of our control.
- Many of these traditional hazards can be mitigated through the purchase of insurance coverage.
- Many insurance packages are in existence today that cover risk events such as credit, political risk, property, trade disruption, and logistics.
- Many of these risk transfer instruments will compensate companies for loss of profits, supplier insolvency, payment default, contract issues, expropriation of assets, property replacement, continuing expenses in the wake of a risk event, and damage to or theft of goods in transit.
- Insurers traditionally utilize a methodology called risk quantification to rigorously evaluate a company's as-is state of risk exposure versus their present amount of insurance coverage in an effort to evaluate the company's risk appetite and then provide insurance recommendations for action to mitigate those risks.

ENDNOTES

1. Adapted from several Reuters articles, Japan Automakers Association, Nissan Manufacturers Association, and the BBC Co., UK.
2. Accessed from Webster's Dictionary.
3. Adapted from the American Insurance Association, accessed from http://www.aiadc.org/AIAdotNET/docHandler.aspx?DocID=287081.
4. Adapted from http://www.zurichcanada.com/internet/can/SiteCollectionDocuments/English/Contingent-Business-Interruption-Canada.pdf.
5. Adapted from AON CSCMP Roundtable Discussion, 2012, and *Supply Chain Management Review* blog with Linda Kornfield, Supply Chain Insurance, 2013.
6. Accessed from http://dictionary.cambridge.org/us/dictionary/business-english/customs-bond.
7. Holbrook, Emily. "Underinsured Nations Exposed to Disasters." *RIMS Magazine*, March 2013.

6

Financial Risk

If we listen to the "experts," we might be tempted to believe that anything that is not strategic to a business must be outsourced. After all, relying on third-party specialists to support the noncore parts of your supply chain is surely better than not relying on specialists. Specialists bring knowledge and skills that are difficult to match internally, skills that should result in lower costs and reduced risk. What could possibly be wrong with that logic?

That is probably what a U.S. manufacturer was thinking when it decided to outsource its freight payments to a third-party payment vendor. Freight payment is the process of applying a set of business rules and fiscal controls around the authorization and payment of invoices for transportation services.[1] The proper use of a third-party expert should reduce a company's risk as the freight vendor ensures accurate and timely payments are made to carriers.

Unfortunately, the payment vendor in this saga began to experience serious financial problems that the manufacturer failed to recognize. As the manufacturer forwarded consolidated payments to the freight vendor for disbursement to transportation carriers, the payment vendor failed to process those payments. Instead, the freight vendor used that money to help meet its own cash flow obligations. It is not hard to imagine what ensued once the financial improprieties became known. Carriers became angry due to a lack of payment, and the manufacturer entered what is best described as a crisis state. Some carriers even started to send collections people after the manufacturing company! Confidence was not restored when representatives from the manufacturer visited the freight payment vendor's office and found a For Sale sign in front of it.

Welcome to the world of financial risk management, perhaps the most thought about risk category since the financial meltdown of 2008. This chapter addresses financial risk by first discussing supplier financial viability and supply market volatility. The majority of this chapter presents approaches for addressing supply chain financial risk, including assessing supplier health using financial ratios, bankruptcy predictors, and qualitative risk indicators; assessing customer creditworthiness; financial hedging; and approaches for managing currency risk. While assessment of financial risk is admittedly not the most exciting topic in the world, it is a major part of the risk management process.

UNDERSTANDING FINANCIAL RISK

Virtually all supply chain risk events have financial implications. Everything that happens within a supply chain eventually ends up on the income statement, balance sheet, or cash flow statement. Our interest in this chapter is concerned with the kinds of events where the primary and immediate effect is financially related. In other words, financial impact is the primary rather than subsequent effect. Two major areas that comprise financial risk include supplier financial viability and supply market volatility, which the following discusses.

Supplier and Customer Financial Viability

Some observers will attribute the growth in the awareness of supply chain risk management to the terrorist attacks on the United States in 2001. However, even though these catastrophic events raised awareness about U.S. vulnerabilities, it was not until the financial meltdown of 2008 that the shock came to the U.S. system, and even to the world, in terms of supply chain risk awareness. At no point since the Great Depression did the financial system of the world's largest economy risk collapsing to the extent that we witnessed in 2008.

Extensive research and experience with companies has helped us reach several conclusions regarding supplier and customer financial risk analysis. Financial risk analysis is about as far as most companies have progressed in terms of managing supply chain risk. Unfortunately, financial risk is not the only kind of risk present in supply chains. Second, given that third-party data about suppliers and customers is increasingly available, it should come as no surprise that most companies begin their supply chain risk management journey looking at financial risk of entities within the supply chain. Keep in mind that assessing financial strength is necessary but is not a sufficient enough part of supply chain risk management to be the only thing being assessed.

Supply Market Volatility

Another major source of financial risk is the volatility of commodity supply markets. The bottom line is that businesses do not like volatility. It makes their ability to plan difficult, if not impossible. If you believe that commodity price fluctuations have increased over the last decade or so, you are right. The size of fluctuations in commodity prices has more than tripled since 2005 compared with the period 1980–2005, based on International Monetary Fund data. Those who follow supply markets know that when demand exceeds supply, the results are allocation of supply, a shifting of power from buyers to sellers, and of course financial risk due to higher prices. Pricing volatility is a direct cause of increased financial risk.

Why do we have such serious commodity fluctuations today? Unlike previous waves of volatility, the current period of fluctuating commodity prices is not driven by a fundamental crisis such as a world war or severe depression. The volatility appears to be a structural change in the way the global economy has organized itself as only eight countries produce the majority of the world's commodities. As demand keeps rising, prices are prone to fluctuations—and this, rather than outright scarcity, is the major challenge. Other challenges facing commodity markets include a willingness of countries (think China here) to manipulate the supply of certain resources, water scarcity, climate change, and energy constraints that limit output. Mining projects in Chile and Mongolia, for example, have been delayed due to energy and water shortages, potentially affecting world prices. Nationalization of commodity companies and the confiscation of foreign-owned assets (such as in Venezuela) are also factors in this era of fluctuating prices.[2] It has also become evident that industrial buyers are not only competing with other companies for commodities today, but they are competing with investors who are increasingly looking to commodity markets for financial returns.

We can also have volatility simply because abrupt changes or shocks occur in the demand or supply side of the commodity equation. We are all familiar with stories involving a facility that explodes and takes with it a disproportionate amount of the world's supply of whatever it is that facility produces. Most of the time we cannot even pronounce what that facility makes. We quickly come to realize, however, the item that just disappeared is a key ingredient that many industries rely on to make their products. Recall from Chapter 1 the fire at a German facility that produces an item called Nylon 12. It quickly became evident after the fire that almost all automotive companies use Nylon 12 and that facility produced a major portion of the world's output. Before that explosion, most of us could not even spell Nylon 12.

Abrupt shifts can also affect the supply side of commodity markets. When a large player or even an entire industry enters a commodity market, sometimes for the first time, the result is usually a dramatic shift in the demand curve with a lagging shift in the supply curve. The inevitable result of this scenario is higher commodity prices. When McDonald's introduced Chicken McNuggets the poultry market became volatile as the demand for chickens quickly exceeded the supply of chickens. Simply put, the chickens needed some time to catch up with this dramatic demand shift (although our sources tell us the chickens put forth a valiant effort). Supply can only increase at a certain rate and usually becomes available in "chunks." The following example illustrates what happens when an entire industry shifts to a commodity for the first time. The results are usually not pleasant.

A Case Study of Supply Market Volatility

Perhaps one of the best case examples of commodity volatility that creates financial and eventually strategic risk involves a relatively unknown element called palladium. This is an element with the chemical symbol "Pd" and an atomic number of 46. It is a rare and lustrous silvery white metal discovered in 1803 by William Hyde Wollaston. More than half of the demand for palladium comes from automotive catalytic converters, which convert harmful exhaust into less-harmful substances. It is also used in electronics, dentistry, medicine, hydrogen purification, chemical applications, groundwater treatment, and jewelry. Palladium also plays a key role in the technology used to develop fuel cells that combine hydrogen and oxygen to produce electricity, heat, and water.[3]

For years auto companies used platinum for emission control in catalytic converters. In anticipation of tougher emission rules by the U.S. government, auto companies began to design palladium into their pollution control equipment. Much to the dismay of other industries that use palladium, the automotive industry suddenly comprised more than half the world's palladium demand. Besides being more effective at cleaning vehicle exhaust, it was also less expensive, at least at the time, than platinum.[4] The end of the Cold War left a large stockpile of inexpensive palladium, particularly in Russia. (Palladium has military applications.) Unfortunately, when an entire industry the size of the automotive industry makes a common change to a new material, one should not be surprised at the resulting shock to the supply market.

As engineers replaced platinum with palladium, the worldwide price of palladium became very volatile. As demand accelerated, the price peaked at almost $1,100 an ounce in 2002, with each vehicle requiring almost one ounce of palladium. As a comparison, 10 years earlier palladium was less than $100 per ounce. At this point macroeconomics would suggest that high commodity prices would encourage suppliers to provide more output, helping to bring supply and demand closer to market equilibrium. Unfortunately, a move toward equilibrium did not happen. First, the primary supply sources of palladium are Russia and Africa, two areas that would not be anyone's top choices as sourcing locations from a risk management perspective. The Russian government, treating the size of its palladium stockpile as a state secret, showed a willingness to "delay" the release of palladium from its stockpile, thereby creating major supply and price disruptions. Also, Western companies were having trouble attracting skilled workers to Africa due to violence and the AIDS epidemic. This had a negative effect on that supply market.

Palladium is also somewhat unusual geologically compared with most other elements. It is mined with platinum in Russia and with nickel in Africa, and the amount of palladium in the ground is proportionally less than platinum or nickel. Some would say it is a by-product of mining other elements. Producers would have to increase their platinum and nickel production to increase their output of palladium, which would cause the prices of those two elements to plunge if demand for nickel and platinum remain steady. While the financial meltdown of 2008 caused the per ounce price to fall to almost $200, the price per ounce doubled in 2010 from around $400 to almost $800 per ounce. Over the last several years the price has fluctuated between $500 and $800 per ounce.

GETTING SERIOUS ABOUT MANAGING FINANCIAL RISK

A variety of approaches exist for addressing financial risk across the supply chain. This is not surprising since corporate insurance, treasury, and investment groups have long thought about risk from a financial perspective. The following approaches for managing supply chain risk borrow heavily from the finance side of the business.

Supplier Financial Health Assessment through Ratio Analysis

A common approach for evaluating a company's financial situation involves ratio analysis. Ratios, also called financial performance metrics, simply represent one number divided by another to provide a value that is then compared to an industry benchmark, internal historical performance, or to other companies, usually in the same industry.

The reasons for calculating and then interpreting financial ratios are straightforward. We use supplier financial ratios to manage risk by providing insights that financial data simply cannot provide. The ratios take financial data and turn that data into value-added information that allows for relatively easy interpretation. Furthermore, ratio analysis, when performed on a regular basis, can highlight trends that can be positive or negative. Ratios can also be used to determine the relative financial strength of a supplier compared with other suppliers in an industry. Perhaps most importantly, various tools use financial ratios to predict the potential of supplier bankruptcy. The most common bankruptcy predictor, the Z-Score, is discussed later. Bankruptcy predictors are actually part of something called predictive analytics, which Chapter 11 addresses.

Ratios are used at different times. Within the context of supply chain risk management, financial ratios are used when—

- Evaluating potential suppliers
- A purchase requirement involves a significant amount of dollars
- Buying items that are critical to the functioning of your business or product
- Entering into a longer-term contractual agreement
- Conducting regular risk scans of your supply chain
- Planning to do business with a supplier where switching options are limited when a company starts using that supplier

Financial Ratio Categories. Literally hundreds of financial ratios exist. Toss some financial data into a numerator and some into a dominator, add a bit of pixie dust, and be prepared to be amazed as a financial ratio magically appears. The first test of a ratio should be that it tells us something of importance rather than simply being the result of numbers thrown into a formula.

Even though hundreds of ratios exist, they generally fall into one of six major categories. It is important to note that not everyone agrees on these categories. Some sources present a different mix of categories, sometimes omitting the market and growth categories presented here and adding ratio categories with names such as solvency, financial efficiency, cash flow, and investment valuation ratios. Interestingly, a search of financial resources reveals that while overlap exists across some ratio categories, total overlap or agreement is rare, if nonexistent. Regardless of the categories used, each ratio category should answer a specific question or satisfy a specific objective unique to that category.

Figure 6.1 presents four ratio categories along with examples of specific ratios within each category. One major category includes *liquidity ratios*, which help identify if a supplier is capable of meeting its short-term financial obligations. A second major category, *activity ratios*, includes ratios that probe how effectively a supplier is managing its assets. A third

Liquidity	Preferred Direction:
• Current ratio: current assets − current liabilities	Higher
• Quick ratio: (current assets − inventories)/current liabilities	Higher
• Cash ratio: cash/current liabilities	Higher
Activity	
• Asset turnover: sales/total assets	Higher
• Current asset turnover: sales/current assets	Higher
• Inventory turnover: sales/inventory	Higher
• Inventory days outstanding: 365/inventory turnover	Lower
Leverage	
• Debt to equity: total liabilities/equity	Lower
• Current debt to equity: current liabilities/equity	Lower
• Interest coverage: earnings before interest and tax/interest	Higher
Profitability	
• Net margin: net income/sales	Higher
• Gross margin: (sales − cost of goods sold)/sales	Higher
• Operating margin: operating income/sales	Higher
• Return on assets: net income/total assets	Higher
• Return on equity: net income/equity	Higher

FIGURE 6.1
Examples of financial ratios.

category is *leverage ratios*. Ratios in this family evaluate whether the supplier is capable of meeting its debt obligations.

A popular and widely used group of ratios that almost everyone agrees on is *profitability ratios*. Ratios in this group provide insight into the rate of return a supplier is earning. Another category that doesn't appear in Figure 6.1 is *market ratios*, a set of ratios that indicate how well a supplier is doing compared with market indicators such as price/earnings and shareholder return. A final category, *growth ratios* (not shown in figure), is somewhat different. Growth ratios provide insight into the rate of growth over time that is occurring at a supplier. These ratios are often compared from one period to another period and require data from multiple periods for their calculation, which is not true of the other ratio categories. Taken at a specific point in time, growth ratios will not provide much insight.

Obtaining Financial Data. Without question the key to successful financial ratio analysis is obtaining reliable data on a timely basis, which is easier said than done. One challenge is that many companies use suppliers that are private companies. These companies are under no obligation to make available the same types of financial documents as required by public companies. Second, the growth in global supply chains means greater use of international suppliers. It is common knowledge that financial data in certain parts of the world might not be accurate or accessible. The method of using two sets of books, which is unthinkable (and illegal) in developed countries is not so unthinkable in other countries. Even data about public companies can be problematic. Large companies in particular will have several if not many operating units and facilities that are aggregated into the financial statements. Breaking out specific results can be challenging, if not impossible.

In practice, companies should establish a repository that contains information from a variety of sources. Internal users should then have easy access to this information when they monitor supplier risk. What are some sources of supplier financial information? A partial listing includes the following:

- Company-published annual reports
- Company-supplied 10-K and 10-Q reports
- Dun & Bradstreet reports
- Credit reports and bank references

- Third-party ratings with an independent firm such as Moody's
- Trade and business journals
- Supplier-provided data

This last item is particularly important. One advantage of working with suppliers on a longer-term basis is the opportunity to share valuable information. Furthermore, as relationships and trust evolve, a supplier should be more willing to share information with a customer, including insights into financial issues that might otherwise not be shared for fear of what the customer might do with that information.

Bankruptcy Predictors

With a crisis often comes opportunity, and that is exactly what happened after the financial meltdown of 2008. A clear need presented itself for ways to assess the financial health of companies that combined sophisticated algorithms and financial data. As a result we are witnessing a growing number of more third-party providers offering sophisticated tools for assessing company health.

One of the most popular and well-established tools for assessing financial health using ratios is the Altman Z-Score. Developed by Dr. Edward Altman of New York University, the Z-Score combines a series of weighted ratios for public and private firms to predict financial bankruptcy. The Z-Score is almost 90% accurate in predicting bankruptcy one year in advance and 75% accurate in predicting bankruptcy two years in advance.

The Z-Score has two attributes that make it a valuable tool for supply chain managers. The first is its relative simplicity. Only four ratios are needed to calculate the Z-Score for private firms and five ratios for public firms. Furthermore, the Z-Score can be interpreted with an easy-to-use red, yellow, and green system. The Z-Score also provides a single score that can be used during the preliminary evaluation of potential suppliers. It provides guidance regarding which suppliers to keep or eliminate from the selection pool. And, it provides a basis for tracking financial changes over time. Supply chain risk managers should calculate supplier Z-Scores at least quarterly. The following are the Z-Score formulas for private and public firms. Notice that a company's total assets appear in three out of four ratios for a private company and four out of five ratios for a public company.

Private Company

$$Z\text{-Score} = 6.56 \times \frac{\text{Working Capital}}{\text{Total Assets}} + 3.36 \times \frac{\text{Retained Earnings}}{\text{Total Assets}} +$$

$$6.72 \times \frac{\text{EBIT}}{\text{Total Assets}} + 1.05 \times \frac{\text{Net Worth}}{\text{Total Liability}}$$

EBIT   Earnings before interest and taxes

where:

Z-Score < 1.1   Red Zone—Supplier is financially at risk

Z-Score between 1.1 and 2.6   Yellow Zone—Some area of financial concern

Z-Score > 2.6   Green Zone—Supplier is financially sound

Public Company

$$Z\text{-Score} = 1.2 \times \frac{\text{Working Capital}}{\text{Total Assets}} + 1.4 \times \frac{\text{Retained Earnings}}{\text{Total Assets}} +$$

$$3.3 \times \frac{\text{EBIT}}{\text{Total Assets}} + 0.6 \times \frac{\text{Net Worth}}{\text{Total Liability}} + 1.9 \times \frac{\text{Net Sales}}{\text{Total Assets}}$$

EBIT   Earnings before interest and taxes

where:

Z-Score < 1.8   Red Zone—Supplier is financially at risk

Z-Score between 1.8 and 3.0   Yellow Zone—Some area of financial concern

Z-Score > 3.0   Green Zone—Supplier is financially sound

The Z-Score Illustrated. It is beneficial to not only present the Z-Score methodology but also to illustrate its use. Table 6.1 provides data for three suppliers that are competing for a contract to provide a base chemical to a pharmaceutical company. For Z-Score calculation purposes the required data for the ratios appears in either the balance sheet or income statement.

TABLE 6.1

Selected Supplier Balance Sheet Data (US$ in millions) for Period Ending
December 31, 2014

	Ninaka Materials	FASE Chemicals	DMS NV
Assets			
Cash	$95.9	$35	$54.3
Marketable securities	$122.5	$9	$27.7
Accounts receivable	$889	$45	$174.5
Inventories	$1057.7	$75	$135.4
Total current assets	$2,165.1	$164	$391.9
Investments at equity	$738.4	$21	$95
Goodwill	$300	$40	$80.4
Total investments and other assets	$1,038.4	$61	$175.4
Property, plant, and equipment	$1,734.5	$125	$412.5
TOTAL ASSETS	$4,938	$350	$979.8
Liabilities And Shareholders' Equity			
Notes payable	$525.5	$11	$35
Accounts payable	$525.9	$75	$125
Taxes due on income	$245	$23	$48
Accrued payroll and employee benefits	$484.2	$13.5	$139
Total current liabilities	$1,780.6	$122.5	$347
Long-term debt	$1,243.5	$55	$165
Common stock	$300	$30	$57.8
Retained earnings	$1,613.9	$142.5	$410
Shareholders' equity	$1,913.9	$172.5	$467.8
TOTAL LIABILITIES AND SHAREHOLDERS' EQUITY	$4,938	$350	$979.8

Statement of Income Data (US$ in millions) Year Ended December 31, 2014

	Ninaka Materials	FASE Chemicals	DMS NV
Net sales	$6,500	$550	$1,355
Cost of goods sold	$5,500	$407.5	$948.5
Selling, general, and administrative expenses	$475	$65	$250
Interest expense	$300	$12	$55
Costs and expenses	$6,275	$484.5	$1,253.5
Income before income taxes	$225	$65.5	$101.5
Estimated taxes on income	$100	$28	$35
NET INCOME	$125	$37.5	$66.5

The net worth figure is also called *shareholders equity* in the balance sheet for publicly traded companies.[5] From an accounting standpoint, *working capital* is defined as the difference between current assets and current liabilities. This figure could be negative if a company's current liabilities exceed its current assets. Earnings before interest and taxes (EBIT) could also be negative in the calculation. The following illustrates the Z-Score for the suppliers presented in Table 6.1:

Supplier Name:	Ninaka	FASE	DMV
Working capital/Total assets × 1.2 =	0.093	0.142	0.055
Retained earnings/Total assets × 1.4 =	0.457	0.569	0.586
EBIT/Total assets × 3.3 =	0.351	0.730	0.527
Net worth/Total liabilities × 0.6 =	0.379	0.583	0.548
Sales/Total assets × 1.0 =	1.32	1.57	1.38
Z-Score =	**2.60**	**3.59**	**3.10**

The following presents the figures used for the Ninaka calculation as a guide:

- Working capital/Total assets × 1.2 = (($2,165.1 current assets – $1,780.6 current liabilities)/($4,938 total assets) × 1.2) = 0.093
- Retained earnings/Total assets × 1.4 = (($1,613.9 retained earnings)/ ($4,938 total assets) × 1.4) = 0.457
- EBIT/Total assets × 3.3 = (($6,500 sales – $5,500 cost of goods sold – $475 SGA expenses)/($4,938 total assets) × 3.3) = 0.351
- Net worth/Total liabilities × 0.6 = (($1,913.9 Shareholders' equity)/ ($3,024 total liabilities) × 0.6) = 0.379
- Net sales/Total assets × 1.0 = (($6,500 net sales)/($4,938 total assets) × 1.0) = 1.32

From this analysis we conclude that only one of the three suppliers Z-Score is in the green zone, indicating the supplier is financially sound. The other two suppliers, while well above the 1.8 Z-Score threshold value that indicates serious financial risk, have some area of concern. For Ninaka the net worth/total assets value is relatively low, while DMV had higher current liabilities compared with current assets and a lower EBIT compared to total assets. Ideally we would like to have financial data for multiple periods so the Z-Score can be compared against other periods. Trend data almost always provides valuable insights that point-in-time data cannot.

Keep in mind that the Z-Score is a financial risk indicator; it does not tell us if the supplier has adequate capacity or can meet quality or delivery requirements. Chapter 13 will illustrate a technique for performing a rough-cut supplier capacity analysis using a small set of the financial data presented in Table 6.1.

Other Predictive Indicators. It is not reasonable to assume that every company has the resources or expertise to evaluate hundreds or even thousands of suppliers from a financial risk perspective. It is also not safe to assume that financial data about these suppliers will be available or easy to obtain. For a variety of reasons buyers will rely on third-party data to support their financial assessment efforts. Two of the more popular predictive indicators are Rapid Ratings and Dun & Bradstreet.

Rapid Ratings (www.rapidratings.com) provides a comprehensive financial health assessment of companies. The assessment is a multiple-part report that provides extensive financial data and analysis. The first page of the output report is a detailed discussion of a company's Financial Health Rating (FHR), which is a number that ranges from 0 to 100. This also includes multiple years of FHR trend data for comparison purposes. The subsequent detailed report contains six sections: Section 1 is the executive summary, Section 2 includes a Return on Capital Employed (ROCE) analysis, Section 3 reports FHR history and performance category scores, Section 4 identifies areas of strength and weakness in relation to other sector participants, Section 5 is the company's balance sheet, and Section 6 is the company's income statement.

Dun & Bradstreet (www.dnb.com) offers a suite of risk management products offering core and add-on modules under a broader category called Supplier Risk Manager. Two analytic tools are available that use predictive scores, including the Supplier Stability Indicator (SSI), a predictor of near term (90–120 days) financial and operational stability and the Supplier Evaluation Risk (SER) Rating, which predicts the likelihood that a company will obtain legal relief from creditors or cease operations without paying creditors in full over the next 12 months.

Qualitative Supplier Financial Risk Indicators

The financial analysis of suppliers should benefit from a combination of quantitative and qualitative assessments. While ratio analysis is a powerful tool, the technique still relies on data that are updated infrequently, difficult to obtain, or sometimes unreliable. The following presents a checklist that might provide clues that a supplier is struggling from a financial perspective[6]:

- A large portion of a supplier's sales go to customers in depressed industries
- A supplier cannot meet agreed-upon lead times because of late purchase order placement for its materials
- A supplier is shipping early due to a lack of business
- The supplier announces facility shutdowns, closings, and/or layoffs
- The supplier has reduced its investment in R&D, IT, equipment, or resources
- Unusual turnover occurs at the executive level
- The supplier's payables period is lengthening
- Quality is deteriorating
- Additional discounts are offered for early payment or payments are required in advance
- The supplier is restating financial reports and projections
- The supplier's product is labor intensive, requiring large payrolls
- The supplier has absorbed up-front research and development and tooling costs on new products that are delayed in getting to the marketplace
- An unusual amount of company stock is being sold by executives
- Rumors of problems begin to emerge

While qualitative indicators are not modeled quantitatively like ratios, they can still provide valuable insights that financial data and ratio analysis cannot. Like forecasting, which benefits from a combination of quantitative and qualitative approaches, assessments of supplier financial health should also benefit from a quantitative and qualitative approach.

Assessment of Customer Creditworthiness

While one end of the supply chain thinks about the financial strength of suppliers, the other end thinks about the financial strength of customers. Effective financial risk management considers both ends of the supply chain. Just as assessments of supplier financial viability should routinely occur, so too should assessments of customer creditworthiness.

The discussion that took place regarding ratio analysis of suppliers from a risk perspective is much the same for customers except with different objectives. A company evaluates the financial integrity of suppliers with the objective of avoiding future supply disruptions due to financial issues while it evaluates the financial integrity of customers with the objective of ensuring its invoices are paid. The interpretation of the various ratios used does not change because a company is a supplier versus a customer.

Similar to supplier financial reports and bankruptcy predictors, risk managers can use third-party data providers to gain insight regarding customer creditworthiness. An example includes the Debtor Risk Assessment (DRA) score from the Coface Group, a 60-year old company that maintains a database of 50 million companies worldwide. Coface calculates a DRA from 0 to 10, with each score corresponding to a different category of trade risk. The assessment comprises primarily data points based on financial ratings, payment incident ratings, and company identity data, although the DRA can be adjusted for external shocks to the economy, trend behavior, or modifications by senior analyst reviews.

Another example of third-party customer risk assessments includes a suite of three reports available from Dun & Bradstreet. While not specifically endorsing any particular product, these reports offer a wide range of customer information that highlights the kinds of data that are available:

- **Comprehensive Insight Plus Report**. The most comprehensive (and expensive) of the three reports, this report provides a business summary, information about corporate relationships, credit limit recommendations, public filings, financial statements, payment history, D&B ratings, and scores that predict the likelihood of future business failure or late payments.
- **Business Information Report**. This report provides essentially the same information as the first report except it excludes the scores that predict the likelihood of future business failure or late payments.
- **Credit eValuator Plus**. This basic report provides customer payment history, industry payment benchmarks, and credit line recommendations.

Other applications and services are available from D&B that target customer credit risk management. A variety of other sources provide basic credit rating scores that can support the development of a credit and accounts receivable policy for a customer.

Hedging

Hedging is a financial risk technique that has been practiced for many years. What exactly is hedging? Hedging is a risk management strategy used in limiting or offsetting the probability of loss from fluctuations in the prices of commodities, currencies, or securities.[7] In effect, hedging is a transfer of risk without buying insurance policies. Hedging employs various techniques but basically involves taking equal and opposite positions in two different markets. One could make a strong argument that a longer-term supply contract that features fixed pricing offers hedges against financial risk. Hedging and hedges can take a variety of forms.

Hedging is a best demonstrated practice for removing pricing risk. Management must be constantly reminded that hedging is a pricing tool. It is used to manage earnings and may result in "lost opportunities" while preventing terrible surprises. Let's be clear about something important—the motivation behind hedging is risk aversion and not monetary gain. If the objective is financial gain, then we are no longer engaged in hedging. We are engaged in speculation. Hedging (risk aversion) is not speculating (risk taking).

Hedging usually occurs within two broad categories—commodity hedging (primarily metals, petroleum, and agricultural futures) and currency hedging. Futures contracts also cover interest rate and stock index futures. Conceptually, hedging involves the purchase and sale of contracts in a marketplace where a gain in one contract is offset by a loss in another, thereby protecting a specific position.

There are two primary kinds of hedging contracts. The first is called futures exchange contracts. These are traded on commodity or currency spot market exchanges. They are rigid in structure and are usually settled on the third Friday of the contract month. Simply turn to the "Money and Investing" section of *The Wall Street Journal* and find the tables for the values of futures contracts looking out into a specified future date (or go to wsj.com/commodities). The second type of hedging contract is called forward exchange contracts. These are issued by banks and traded among financial institutions. While similar to futures contracts, they are individually contracted and provide more flexibility.

Why should a company engage in hedging? Consider the following scenario where a company decides not to engage in hedging. An apparel company creates a 2016 budget that considers cotton prices at $0.88 per pound that reflects the price in September 2015. The company expects to buy 1 million pounds of cotton in 2016. The 2016 financial plan will build in an expectation of $880,000 (1 million × $0.88) for cotton. Now, let's fast-forward to 2016, when cotton has increased to $1.00 per pound. If $1.00 per pound is the average price over the year, this company will see its operating income drop by $120,000 ($1 million actual 2016 cotton expenditure less $880,000 budgeted cotton expenditure) compared to the budget plan. This company should also expect to see a decline in all its financial returns measures. If this company had purchased a futures contract at $0.88 per pound, it could have avoided the $120,000 "loss" from purchasing the cotton in 2016 prices. Of course, if the price of cotton declined to $0.76 in 2016, the company would have foregone a $120,000 "gain." But, do not forget that the point of hedging is not to gain or lose. It is to provide stability that allows for more effective planning and the avoidance of major surprises.

Options. Buyers can purchase a third type of contract called futures options. Options allow the holder to exercise the future contract at a pre-established price or to walk away from that contract. If a commodity market shifts in favor of a buyer, which likely means the commodity is lower than the futures contract price, the buyer will take advantage of the market shift and forego exercising the contract. If the market shifts against the buyer, which likely means the commodity price has moved higher than the futures contracted price, the buyer will exercise the option at the contracted futures price. As your parents probably said, if something sounds too good to be true, it probably is. The reality is that purchasing options can be quite expensive. This points out once again that in business there is no such thing as a free lunch.

Let's illustrate futures options with a currency example. A buyer orders equipment directly from a German company that costs 10 million euros (€) with expected delivery in six months. The company can purchase a futures option to buy euros in six months at €1 = $1.28. Conversely, $1 = €0.78. The cost of the equipment in U.S. dollars is $12.8 million (€10 million × 1.28 or 10 million / 0.78), which the futures contract locks in.

Now, let's look at two scenarios and predict what would likely happen. In Scenario A, the exchange rate in six months changes to €1 = $1.20, which makes the cost of the equipment in U.S. dollars $12 million (€10 million × 1.2) at time of delivery. In this case the buyer will likely walk away from the futures contract and purchase the euro on the open market, assuming the currency savings are greater than the option cost. In Scenario B, the exchange rate changes in six months to €1 = $1.35, which makes the cost of the equipment in U.S. dollars $13.5 million (€10 million × 1.35) at time of delivery. In this case the buyer will exercise the option to pay with euros at the futures contract rate of €1 = $1.28 and "make" $0.07 per euro.

In closing, hedging is a well established but certainly not foolproof way to mitigate financial risk. The list of items that cannot be hedged is much longer than the list that has active futures markets. This limits where hedging can be applied. Furthermore, supply chain managers are strongly urged to work closely with finance before even thinking about hedging activities. This is one area where specialized finance knowledge is critical. Hedging, options, and commodity and currency trading are not for the faint of heart or the ill-prepared.

Currency Risk Management Approaches

Whenever a company engages in international buying or selling, it raises the possibility of currency fluctuations that present financial risk. This does not always have to be the case. For example, some countries use the U.S. dollar as their currency. These are generally not major countries or trading partners of the United States. Examples of countries that use the dollar are Bermuda, Panama, Ecuador, and El Salvador. The technical term for this is *dollarization*, which occurs when the inhabitants of a country use foreign currency in parallel to or instead of the domestic currency as a store of value, unit of account, and/or medium of exchange within the domestic economy. The term is not only applied to usage of the U.S. dollar, but generally to the use of any foreign currency as the national currency.[8] Currency risk involving countries that use the dollar as their currency is essentially nonexistent. Other countries may use the euro as their currency while others use the Australian or New Zealand dollar, Swiss franc, or Indian rupee.

A second scenario involves countries that freely accept a second currency for transactions in addition to their own currency. There are probably a dozen or so countries that use the U.S. dollar alongside their own currency. This too can have the effect of minimizing currency risk since the dollar is an accepted currency.

A third scenario involves countries that peg their currency against a specific value of a second currency. This scenario also minimizes risk because a country sets or "pegs" its currency to a specific value of another currency. These pegged values are usually stable over time. For many years the Chinese government pegged its currency, the renminbi or yuan, to a specific value of the U.S. dollar. It is only fairly recently that the Chinese government has allowed its currency to trade a bit more freely. Its trading band, however, is still tightly controlled by the Chinese central bank. Currency risk with countries that peg their currency to a company's country currency is fairly small, at least in the short term.

The final scenario, and the one that presents financial risk, occurs when one currency freely floats against the value of another currency. This means the relative value of a currency could increase or decline relative to another currency. While a currency shift could work in a company's favor, it could go also go the other way.

What are some approaches for managing currency risk? The following list, while not exhaustive, shows that a variety of methods can be used in isolation or in combination with other methods to address currency risk issues.

Currency Hedging. The previous section illustrated the use of options and currency hedging, so there is no need to repeat that information here. Without question, hedging is one of the most common ways to address currency risk when engaging in international business.

Purchase in the Home Country Currency. It should come as no surprise that many companies, especially smaller companies, manage currency risk by avoiding it. One way they do this is to write international purchase contracts in the company's home currency. When this happens, the other party assumes the currency risk. How could there possibly be a downside to this approach? If you think this approach sounds too good to be true, you're probably right.

The party that accepts the currency risk knows it has taken on this risk. And it should come as no surprise that this assumption of risk will likely come with a cost. We should expect the party that assumes the risk, particularly suppliers, to add some factor into their unit price to compensate for its risk management costs or exposure.

Offering to assume the currency risk could result in a lower unit cost, which has clear financial implications. Do not forget that inventory is treated as a current asset on the balance sheet. If the value of that asset is being inflated due to currency risk management costs, this will adversely affect a wide range of high-level financial indicators, including return on assets and return on invested capital. More advanced companies try to take out all the "cost adders" that suppliers include so they can better manage these cost elements. As an aside, risk management costs are certainly not the only cost adder or factor that suppliers might include in a purchase price. Suppliers might include adders when customers fail to pay invoices in a reasonable time or constantly change order quantities.

Sharing Currency Risk. The parties to a contract can agree to share risk when currency rates fluctuate. With this approach the responsibility for managing currency risks does not fall to a single party. This is a reasonable approach if the parties to a contract are confident that currency rates will remain relatively stable over a specified period or when dealing with shorter-term contracts. The reality is that most currency rates fluctuate at slower rates compared with commodities.

Currency Forecasting. Currency forecasting, which should occur regularly, can be performed by corporate finance groups, external third parties, or a combination of the two. The results of the forecasting process can affect the degree to which a company engages in currency risk management. If a forecast predicts minimal shifts in the relative value of one currency against another, then a buyer will likely not focus too extensively on currency risk as an issue, at least in the short term. Currency shifts that are forecasted to be more dramatic or uncertain invite more aggressive risk management approaches.

Contract Escape Clauses. A common risk management approach, usually insisted upon by the legal group, is to have language in a purchase contract that allows a party to opt out of an agreement due to currency issues. This generally does not happen when the contract is for a piece of capital equipment that has a longer lead time. Rather, it is more common with contracts that involve repetitive purchases. A supplier would likely be reluctant to enter a contract to build and sell a $5 million piece of equipment with the risk that a buyer can simply walk away from the contract because of a currency shift.

Currency Renegotiation or Adjustment Clauses. Another approach for addressing currency risk is to stipulate when formal currency reviews will occur. These reviews are generally triggered in two ways—delivery-triggered reviews and time-triggered reviews. With delivery-triggered reviews, a supplier delivery triggers the currency review, while time-triggered reviews happen at specific dates.

Most parties recognize that some normal or accepted currency risk takes place when using currency renegotiation or adjustment clauses. Figure 6.2 illustrates the concept of normal currency risk in a time-triggered adjustment clause. As long as the value of the Japanese yen to the dollar stays within an agreed-upon bandwidth, in this case 90–102 yen to $1, no

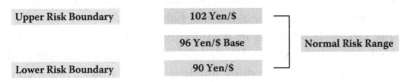

Time-Triggered Adjustment Clauses

An annual contract for electronic components is agreed to with a Japanese producer in January. A currency clause is agreed to that stipulates currency reviews will occur quarterly. The exchange rate at the time of signing the contract is 96 Yen = $1. The parties have agreed that a band of normal currency risks is +/− 6%. Adjustment review dates are April 1, July 1, and October 1.

Upper Risk Boundary	102 Yen/$	
	96 Yen/$ Base	Normal Risk Range
Lower Risk Boundary	90 Yen/$	

FIGURE 6.2
Currency risk management.

formal review or discussion takes place when a review is triggered. This bandwidth represents normal risk that the parties are willing to accept. If, however, the relative currency values move outside the accepted bandwidth, a formal currency review is triggered. At that point various actions might be agreed to by the parties, including currency risk sharing or possibly even the termination of the contract.

Regardless of the approach used to manage currency risk, it is recommended that supply chain professionals work with their counterparts in finance before engaging in any activities that take the supply chain professional out of his or her comfort zone, something that in itself could lead to unnecessary risk.

CONCLUDING THOUGHTS

While financial risk assessment may be more mature than other risk management techniques, this does not mean that challenges do not remain or that further improvements are not possible. The challenge with any supply chain technique, including financial assessment techniques, is moving beyond first-tier suppliers. Second-tier suppliers that suffer a disruption due to financial difficulties will soon affect tier-one suppliers, which will quickly affect the operations of your company. If your executive leadership assumes that there is nothing to worry about because surely your tier-one suppliers take the pulse of their tier-one suppliers (your tier-two suppliers), at least take comfort in knowing there are countless others who share this same delusion.

Summary of Key Points

- Supplier financial risk analysis is about as far as most companies have progressed in terms of managing supply chain risk.
- Given the increasing availability of financial data it should come as no surprise that most companies begin their supply chain risk management journey looking at supplier financial risk.
- The size of fluctuations in commodity prices has more than tripled since 2005 compared with 1980–2005 based on International Monetary Fund data, making supply market volatility a major contributor of financial risk.
- A variety of approaches exist for addressing financial risk across the supply chain, which is not surprising given that risk at the corporate level has historically been defined largely in terms of financial risk.
- The financial analysis of suppliers should benefit from a combination of quantitative and qualitative assessments.
- A common approach for evaluating a company's financial situation involves ratio analysis. Without question the key to successful financial ratio analysis is obtaining reliable data on a timely basis.
- One of the most popular tools for assessing financial health using ratios is the Altman Z-Score. The Z-Score is only a financial risk indicator; it does not tell us if the supplier has adequate capacity or can meet quality or delivery requirements.
- A best practice for removing pricing risk is hedging. Conceptually, hedging involves the purchase and sale of contracts in a marketplace where a gain in one contract is offset by a loss in another, thereby protecting a specific market position.
- Whenever a company engages in international buying or selling, it raises the possibility of currency fluctuations that present financial risk.
- A variety of methods are available to help manage currency risk, including currency hedging, purchasing in the home country currency, sharing currency risk, currency forecasting, contract escape clauses, and currency renegotiation or adjustment clauses in contracts.

ENDNOTES

1. Accessed from http://corporate.traxtech.com/5mistakes.pdf.
2. Eisenhamme, Stephen. "Governments Must Tackle Sharp Commodity Price Swings-Think Tank." *Reuters*, December 12, 2012. Accessed from www.reuters.com.

3. Accessed from www.wikipedia.com.
4. White, G. L. "Unruly Element: Russian Maneuvers Are Making Palladium More Precious Than Ever." *Wall Street Journal*, March 6, 2000: A1.
5. Accessed from http://www.investopedia.com/terms/s/shareholdersequity.asp.
6. "10 Warning Signs of a Supplier in Peril." *Industry Week*, April 2009: 38.
7. Accessed from www.businessdictionary.com.
8. Accessed from http://en.wikipedia.org/wiki/Dollarization.

7

Operational Risk

Imagine a company that sells some of the hottest products on the planet with just a few products representing a disproportionate share of revenues. Imagine further that this company has relied on a single Chinese supplier and location to build these products. And imagine even further that this Chinese supplier is secretive, often showing an unwillingness to share information publicly, particularly about labor problems. If this scenario sounds operationally risky (and strategically risky as well), you just figured out why Apple has made a decision to diversify its supplier base by expanding its outsourcing to a second supplier located in Taiwan.

This chapter will discuss the many potential operational failures through the prism of the four pillars of supply chain risk—supply risk, demand risk, process risk, and environmental risk. We will profile operational risks that happen every day in these four areas. We'll also discuss these risks in the context of two basic operation horizons and present a program that traditionally has been utilized to mitigate and manage operational risks.

OPERATIONAL RISKS

By far, a disproportionate set of supply chain risks will be categorized as operational since this category includes internal and external quality problems, late deliveries anywhere in the supply chain, service failures due to poorly managed inventory, problems related to poor forecasting, and a thousand other events related to operational performance failures. This section discusses the operational risks that are part of supply risk, demand risk, process risk, and environmental risk. The two prevalent horizons affecting supply chain risk management are the operational horizon, covering 0–45 days into the future, and the tactical horizon, which normally covers 1–18 months into the future.

Supply Risk

Referring back to the discussion in Chapter 2 regarding the four pillars of supply chain risk, the supply management profession is by far the most mature discipline within the supply chain arena when it comes to identifying, assessing, mitigating, and managing risk. Procurement professionals have been leveraging techniques to mitigate and manage risk for more than 50 years. Let's briefly identify operational and tactical risks that reside within this risk pillar every day. In no particular order, we have supplier lead times, supplier quality, supplier prices, supplier insolvency and bankruptcy, supplier delivery issues, fraud, corruption, counterfeit parts and components with subtier suppliers, and inbound logistics. To further focus our discussion, we'll classify these risks into supplier, logistics, and fraud, corruption, and counterfeiting. Table 7.1 can be used as a reference throughout the supply risk discussion.

Supplier Risks. As mentioned, procurement professionals have been trained for many years to think about risk and contingencies, probably much more so than any other discipline within the supply chain community. One of the main reasons is most manufacturers' cost of raw material represents approximately 50%–70% of their total cost of goods sold. That's a huge portion of the total cost of finished products and an abnormally large risk element to the organization.

As shown in Table 7.1, the traditional approach to handling supplier risk has been to use buffer inventory or statistically derived safety stock to absorb volume shocks or delays or supplier delivery and quality issues. One of the traditional techniques many procurement professionals have been trained to execute to ensure better pricing and better delivery has been placing more and more of their raw material requirements with one supplier. This traditional thinking and training was driven by the premise that when a company's purchase requirements become a larger portion of a supplier's order board, that supplier will bend in terms of price and do its best to demonstrate solid delivery performance because of the risk of losing those orders and volume.

This procurement strategy worked well in a stable environment before globalization and supply market volatility. What actually took place is many companies got a bit complacent performing their due diligence.

TABLE 7.1

Supply Risks

Supply Risk	Cause	Horizons	Traditional Remedies
Supplier lead times	Material/capacity issues	Both	Buffer stock, larger order quantities
Supplier quality	Manufacturing processes	Both	Contract verbiage, penalty clauses, inbound inspection
Transportation lead time	Breakdowns, acts of God, customs issues	Both	Contract verbiage, penalty clauses
Subcontractor availability	Initial source can't deliver	Both	Contracts for potential capacity reservation
Supplier pricing	Performance issues, contract changes, breach of contract	Tactical	Due diligence, phone-fax, and possible visits
Time delay	Customs, lack of performance	Both	Buffer stock, rescheduling final delivery
Disruption	Labor issues, natural disasters, terror	Both	Buffer stock, safety stock, second source capacity
Import delays	Customs paperwork, port strikes, labor issues	Both	Additional freight forward companies, calls to government contacts
Supplier insolvency	Poor management, acts of God, force majeure	Both	Loans, law suites, litigation, and second sourcing
Fraud/ corruption	Poor government oversight	Both	Fines, penalties, and operating restrictions
Counterfeit material	Poor government oversight	Both	Fines, penalties, and operating restrictions
Supplier delivery	Manufacturing issues, quality issues, customer requirement changes	Both	Buffer stock, warehouse inventory, second source

Buyers assumed a bit too much in terms of continued performance and found out through disruptions, supplier insolvency, supplier quality issues and much more, that relying on one supplier for a major portion of a company's purchase needs might not be the best strategy when it comes to managing risk across an entire organization.

Logistics Risks. Around the year 2000, during the Internet and e-business boom came the concept of Business Process Outsourcing (BPO). IBM was a big proponent of this concept within its own supply chain and promoted doing what you do best and outsourcing the rest. Why did IBM and many other companies embrace this approach to supply chain management?

One reason was companies found that there were many organizations around the globe that could do certain business functions better, faster, and cheaper. And at that time, as the Internet was exploding on the supply chain scene, there was renewed interest in exploiting the World Wide Web to collaborate with these new BPO organizations and new partners to drive overwhelming top-line growth.

With that belief, a new industry called third-party logistics, or 3PLs, was born. A quick story about the chemical industry in 2000 will perhaps provide some context regarding the growth of 3PLs. During this period the chemical industry was still vertically (i.e., functionally) aligned in terms of supply chain management. There were several reasons that many organizations had not embraced the concept of supply chain management. First, many organizations had very good profit margins and did not see the need for change. Second, the "chemists" were still running chemical companies and lacked supply chain knowledge. And third, the chemical industry was still an asset-intensive industry that believed in the benefits of vertical integration.

During this time, a few leading-edge chemical organizations began benchmarking their total logistics costs associated with inbound and outbound material delivery as a percent of total sales. The numbers were alarming. On average, the chemical industry's transportation costs were more than 10% of sales. As the industry benchmarked against other industries, it came to the conclusion that while chemical companies were good at breaking down hydrocarbons, these companies were not so good at logistics. As a result these companies, like so many in other industries, outsourced their logistics to companies that could service their needs at a much reduced rate.

If we scan Table 7.1 and view the logistics risks, we're not saying the BPO approach has totally eliminated logistics risk. However, these outsource providers have developed many tools and techniques to manage risk for their customers. Many of the traditional remedies are still being leveraged but by a new industry that tends to have much more experience in global trade and has invested in more advanced tools and techniques.

Fraud, Corruption, and Counterfeiting Risks. The European Banking Board has developed a set of baseline definitions for their employees and their customers to follow. Fraudulent practice means any action or omission, including misrepresentation, that knowingly or recklessly misleads or attempts to mislead a party to obtain a financial benefit or to avoid an obligation. Corrupt practice means the offering, giving, receiving, or soliciting, directly or indirectly, of anything of value to influence improperly the actions of another party.[1] Counterfeiting occurs when something is made in imitation so as to be passed off fraudulently or deceptively as genuine.[2] This portion of supplier risk is far too large to dig deeper into at this time, but we will talk at length about fraud, corruption, and counterfeiting in Chapter 8.

Demand Risk

Demand management has always been a difficult discipline by definition. Part of this is due to the tendency of forecasts that are almost always wrong to some degree. Demand management and forecasting techniques and solutions have been available to the supply chain profession for over 80 years. There are hundreds of deterministic, statistical solution providers that provide companies with the ability to scan historical sales to arrive at a forecast using techniques such as least squares, time series analysis, and regression analysis. We'll talk in more detail about these techniques in Chapter 10, but for now we'll segment the demand risk discussion into customer risk, product risk, and logistics risk.

Customer Risk. As shown in Table 7.2, there are plenty of risks on the customer side of the equation. The demand issue tends to get the most focus because the purpose of demand estimation is to project what a customer will buy, when they will buy it, and how many they will buy. With complex supply chains and large product portfolios, not even considering global markets, seasonal products and other extraneous factors, the task is somewhat daunting.

Forecast error is a key risk in demand management that requires attention. The reason we focus on this risk, is that it runs from 10% error of forecast versus actual at the aggregated product family level to more than 40% error at the stock-keeping unit or item level. (Forecasts almost always become less reliable as the forecasts become more granular rather than aggregated.)

The traditional remedy to mitigate forecast error has been to statistically calculate safety stock and develop buffer stocks at choke points throughout the supply chain. The supply chain profession has been working for more than 50 years to mitigate the risk of attempting to project what their customers will buy and balancing supply to ensure superior service levels to those customers. And industrial customers are in no way exempt from providing additional risk to this pillar of supply chain risk. Customers

TABLE 7.2

Demand Risks

Demand Risk	Cause	Horizons	Traditional Remedies
Forecast error	Seasonal issues, lead times, poor information, inadequate systems, poor communications, inadequate skills	Both	Statistically derived safety stock, buffer stock points, excess inventory throughout supply chain
Time delays	Customer changes, systems issues, product issues	Both	Rescheduling, price concessions
Outbound transit times	Carrier issues, acts of God, customers' issues	Both	Carrier discussions, customs calls, freight forwarding follow-ups
Customer pricing	Poor communication, inadequate contract verbiage, poor performance	Operational	Concessions, Rescheduling deliveries
Customer promotions	Poor communications, poor execution on both sides	Operational	Constant conference calls, rescheduling manufacturing and deliveries, stealing product form other customers
Customer bankruptcy	Poor execution, fraud, corruption, sell out by owner	Both	Possible loans, possible merger or partnership
Product failure	Poor quality control, material issues, incorrect specifications	Both	Rescheduling, modifying the specifications, price concessions
Warranty issues	Poor communications, poor specification management, recall of product, death and more	Both	Law suites, litigation, government involvement, fines and penalties
Customer loss	All of the above issues and more	Both	Sell off material designated for customer, write off if specific and scan for new customers
New product introduction	Poor planning, poor communication throughout organization, poor execution, poor assumptions	Both	Ad hoc meetings, excessive overtime, price modifications, new promotions, rescheduling of manufacturing plans
Fraud and corruption	Poor government oversight	Both	Fines, penalties, and operating restrictions

have a nasty habit of continually arguing about changes to pricing and delivery dates and have a propensity to surprise suppliers with unplanned or poorly communicated product promotions of the supplier's products at their stores.

When a customer surprises a supplier with a promotion of the supplier's product, the resulting demand tends to be at least three to four times the forecasted level of sales, something that wreaks havoc on the supplier's supply chain. The typical approach to respond to this situation is to reallocate product within the supplier's bill of material to satisfy the promotion. This approach tends to result in excessive overtime at the supplier, total disruption of the supplier's supply chain–planning processes, and aggravation to other customers when their delivery promises are rescheduled. And finally, if this risk is perpetuated, or becomes chronic at a supplier, the outcome tends to be a reduction in customer relationships and ultimately a loss of customers.

Product Risk. Poor product portfolio management is another important aspect of this risk pillar. By far the largest risk in this category is product failure and warranty issues. An example we all have witnessed over the last couple of years is Toyota's issues with braking systems, accelerators, and massive product recalls. Automobile manufacturers run the numbers on their risks associated with product liability and warranty probably better than most manufacturers. Their risk appetite is usually quite high and they utilize many diverse liability, tort, and warranty insurance packages to mitigate those risks. However, continued product recalls, regardless of the industry, can lead to customer loss, fines, penalties, and potential bankruptcy. Witness what has happened at General Motors.

One of the most difficult elements in this pillar is new product introduction. Forecast error for products continually produced and sold to customers can become as large as 40% for a given product. Forecast error and the impact on the company of poorly launched new products can be even more dramatic. We mentioned this aspect of demand management and supply chain management in Chapter 4. Poor assumptions for poor planning and poor communications relative to new product launches can be devastating. An example is the growth of Apple's iPhone and Samsung's Galaxy products at the expense of Blackberry's product portfolio.

There are traditional approaches to mitigate product-planning risks including collaborative planning, forecasting, and replenishment (CPFR) tools that provide a vehicle for demand collaboration between suppliers and customers. CPFR is utilized in the consumer packaged goods and retail industries. The sales & operational planning (S&OP) process acts as a framework to assist companies in their efforts to balance supply with demand to minimize surprises and maximize profits for the entire enterprise. We will discuss these in more detail in Chapters 9 and 12.

Logistics Risk. In this risk pillar, logistics relates to outbound material that perhaps goes to a final assembly/package partner, a distribution or warehouse, or the final customer. Most of the logistical discussion we engaged in within our supplier risk segment applies in this arena as well. Most manufacturers do not maintain their own truck, barge, rail, or ship fleet. As mentioned earlier, thousands of 3PL companies will haul freight better, faster, and cheaper than most manufacturers. One nuance is the approach that utilizing 3PLs to deliver finished goods to customers is a form of risk mitigation due to better performance, or if contract verbiage with that carrier includes sharing or pooling of the risk in case of an accident. In the chemical industry, contracts specify not only what we call service level agreements (SLAs) but also language that shares the liability risk of an accident with both parties. This has been a traditional remedy to improve service deliveries, reduce the cost of transportation, maximize profits, and minimize risk.

Process Risk

In this risk pillar, which a cursory glance at Table 7.3 reveals to be extensive, the risks are inherently positioned within an organization. Another way to think about this is that the organization has better control of these risks because they occur within their own domain. The frequency of occurrence and the remedies many organizations utilize to solve these issues lie within their own four walls. Our categories for this risk pillar discussion will be known or hard risks, unknown or soft risks, and chronic risks that can arise within a company's four walls.

Known Risks. These are risks that are measurable and can be planned for. Known risks, also called hard risks, include process breakdowns, poor material, poor quality control, criminal activity, poor and unreliable systems, and failure of a company's facilities and assets. Known risks from Table 7.3 could include manufacturing yield, capacity, information delays, systems, receivables, payables, inventory, and planning. Most of

TABLE 7.3

Process Risks

Process Risk	Cause	Horizons	Traditional Remedies
Manufacturing yield	Equipment failure, material issues, human error	Operational	Reschedule run, cut into existing capacity plan
Capacity	Equipment failure, poor performance, poor communications, poor planning	Both	Reschedule runs, reschedule deliveries, possible use of contractors
Information delays	Poor planning, inadequate systems, outages	Both	Backup systems, ad hoc meetings, extreme overtime
Time delays	All of the above and below	Both	Ad hoc meetings and excessive overtime
Disruption	Labor issues, systems, material, inbound material, natural disaster or act of God	Both	Ad hoc meetings and excessive overtime
Systems	Outages, terror, hackers, internal errors	Both	Backup systems, ad hoc meetings, vendor outreach and excessive overtime
Receivables	Poor execution, poor contract verbiage, poor relationships, customer financial issues	Both	Phone calls, e-mails, faxes, visits and possible collection agencies
Payables	Cash flow issues in-house, cash flow strategy, poor relationships with suppliers, poor supplier performance	Both	Phone calls, e-mails, visits and possible contract renegotiations
Inventory	Forecast error, product life cycles, poor planning systems, poor supply chain management execution	Both	Excessive safety stock, write-downs and write-offs
Intellectual property	Outsourcing, contractors, partnerships, and espionage	Both	Vertical integration, contract verbiage, fines and penalties

continued

TABLE 7.3 (continued)

Process Risks

Process Risk	Cause	Horizons	Traditional Remedies
Human/ process error	Operator issues, fraud, corruption, systems breach	Both	Process revalidation, employee reeducation, law suites/or discharge
Planning	Inadequate systems, inadequate training, poor supervision, poor management style	Both	Systems upgrades, reeducation, additional collaboration, and metrics of success
Product failure	Poor material, poor quality control, poor communication, poor management oversight	Both	Enhance communications, customer visits, supplier visits, contract renegotiations
Equipment failure	Poor maintenance schedules, operator error, material issues, component failure	Both	Perform assessment, revalidation of alternative equipment/routings, vendor visits in-house
Organizational management	Poor performance, poor communication, inadequate measurements, fraud	Both	Ad hoc meetings, assessment of skill sets and possibly enhancement of roles, goals and measurements
Strategy	Poor planning, poor execution, poor communication, competition	Both	Same as above with possible change in strategy

these risks are known, albeit not particularly popular, because they are indicative of a management style and attitude or a lack of investment of time and resources. Many of the risk remedies are reactive and many of the risks, without proper attention can become chronic, which we'll discuss shortly.

The critical issue with known risks is that without addressing them with permanent solutions, an organization tends to be plagued with these risks daily to the point where they lose visibility, confidence, and commitment, leading to a potential major internal disruption. This breakdown within a company's processes can be as impactful as a catastrophic event emanating externally.

Unknown Risks. These risks are difficult to determine and are sometimes called soft risks. Examples of soft risks might be a radically new product or technology that renders a company's existing approach to the market obsolete. Unknown risks could also be a fire that destroys a plant, an attack on a plant, a weather event, and from Table 7.3, areas such as time delays or any unforeseen disruption. An effective way to respond to these risks is to develop and practice response scenarios, what we call business continuity planning (BCP). The disconcerting aspect of soft risks is that most companies do not develop or practice scenario response plans or risk response plans. In fact, as you scan down the traditional remedies column for the soft risks in Table 7.3, you will notice that almost all are reactive rather than proactive.

Chronic Risks. The primary characteristic of chronic risks is that when these occur they tend to cause only minor internal disruptions. They may occur continually and because of the nature of their low impact, organizations tend to absorb the risk and develop work-arounds. The disruptions could be persistent and the root causes may not be obvious and therefore become tolerated over time. Some of the risks from Table 7.3 that could fall into this category include manufacturing yield, capacity issues, time delays, human errors, and equipment failure.

Environment/Ecosystems Risk

The fourth risk pillar is probably the most immature pillar since there are so many new government rules and regulations, weather events, and fraud and corruption possibilities emerging around the globe. Furthermore, most organizations are operating global supply chains in areas where they've never operated before. This is all in an effort to grow top-line revenues and penetrate new markets. Globalization strategies bring additional risk, which Table 7.4 profiles.

Known Risks. In this arena we could categorize risks such as currency rates, customs regulations, environmental regulations, industry regulations, and country regulations. We may not like all the regulatory statutes placed upon us, but they tend to be known and developed over a wide time span, thus providing organizations ample time to prepare for and comply with these rules. Many companies do not have the skill sets to understand and manage all the rules and regulations and therefore rely on 3PLs and freight forwarders to ensure compliance. One caveat before we move to the unknown risks is that all companies have a distinct style and attitude regarding risk, and sometimes their risk appetite is not what it should be. Subsequently, they may or may not choose to adhere to all the rules.

TABLE 7.4

Environment/Ecosystem Risks

Environment/ Ecosystem Risk	Cause	Horizons	Traditional Remedies
Currency exchange rates	Central banks, country issues, conflicts	Both	Use of financial hedging techniques
Political environment	Conflicts, political upheaval	Both	Calls with country officials, tapping own government contacts
Customs regulations	Improper paperwork, poorly packaged material, terror	Both	Use of 3rd party logistics partners, conversations with customs, enhanced paperwork
Weather/acts of God	Floods, tornados, hurricanes, fires, volcanoes, war	Both	Disaster insurance
Environmental regulations	Lack of discipline, failure of audit, poor management and diligence	Both	Excessive overtime for remedial compliance
Industry regulations	Same as above	Both	Same as above
Country regulations	Same as above	Both	Same as above
Fraud/ corruption	Country policies or lack-thereof, suspect partners, misrepresentation by 3rd party contractors	Both	Fines, penalties, shutdowns and remedial policy enhancements, including discharge
Counterfeiting	Same as above	Both	Same as above, including alternative sourcing and partnerships
Competition	Lack of focus, poor company communication, poor product introduction process, poor execution	Both	Price reductions, marketing promotions, customer visits, enhanced product portfolio and extended warranties

Unknown risks. Risks within this category could be political, weather and acts of God, fraud, corruption, counterfeiting, and competition. The bulk of these risks is mitigated and managed through the use of scenario-based planning approaches, be it at a specific facility level or throughout an entire supply chain network. Most of the unknown or soft risks in this category can and should be planned for using scenario-based BCP or risk response plans. However, a few of these risks, such as fraud, corruption, theft, and counterfeiting lend themselves to a more reactionary approach.

Historically, almost every soft risk is event-driven in nature. Theft has traditionally been a reactionary risk, followed by countermeasures, incident follow-ups, and bulletins to recover the loss. Counterfeiting has traditionally been much more of a sensitive subject, because based on the industry, when it occurs governments tend to get involved to protect citizens. And fraud and corruption is a sensitive subject since no company wants its brand on the news or the web for the wrong reasons.

When these soft risks emerge, whether inside or outside a company, organizations should address the issue as quickly as possible. Many fraud events emanate from within the organization and are dealt with quietly using third-party fraud investigative companies. When fraud, bribery, and corruption emerge outside the organization, in countries where a company's products are being manufactured or sold, that's when the brand is most at risk. The bottom line in this area of risk is that it will continue to grow in terms of scope and scale as long as companies continue to penetrate new global markets. The approaches to manage these risks will also continue to emerge and perhaps migrate from reactive to more proactive in nature. We'll share many of those new proactive approaches in subsequent chapters.

BUSINESS CONTINUITY PLANNING

Adopting a business continuity plan (BCP) is the start of a journey that ensures continuous operations of critical processes within a company and expands to include critical suppliers as the program matures. It is a concept that is absolutely central to effective risk management. In reality, this topic could appear in one of many chapters. We simply made a decision to place business continuity planning in this chapter because so many supply chain risks are operational in nature. Before describing business continuity planning in detail, we will define some important concepts and definitions.

Business continuity is the process of planning for and implementing procedures that are designed to enable continuous operations of critical business processes and functions. Incident management is the process that is responsible to guide the company through an incident or disaster and execute the overall business continuity plan. The incident management team focuses on incidents that have escalated beyond emergency response and that could impact business operations (i.e., business continuity). The responsibilities of the incident management team include the following:

- Activate department business continuity plans and disaster recovery plans as appropriate
- Make workplace recovery decisions
- Activate disaster recovery decisions
- Allocate resources among recovering departments/groups
- Coordinate efforts between recovery and response teams
- Approve disaster-related purchases
- Develop and distribute messages to employees, customers, and vendors
- Provide direct updates to the executive team
- Carry out governance board and executive directives

Emergency response is the process that is responsible for human and life safety issues during an incident. The emergency response team leads the evacuation and assembly or shelter-in-place activities.

BUSINESS CONTINUITY PLANNING OBJECTIVE[5]

The objective of a business continuity plan is to ensure the availability, reliability, and recoverability of business processes servicing a company's customers, partners, and stakeholders. In order for business continuity to be effective, it must be an integral part of the business planning life cycle. Whenever business changes impact a process or function, business continuity considerations must be evaluated and adjusted as necessary to understand the effect to existing recovery strategies and plans. We all make plans based on trade-offs of cost and benefits. Business continuity formalizes a company's overall approach to effective risk management and should be closely aligned to a company's incident management, emergency response management, and information technology disaster recovery. Successful business continuity management requires a commitment from the company's executive team in order to show commitment, raise awareness, and implement sound approaches to build resilience.

The Business Continuity Life Cycle

The business continuity life cycle includes six stages:

1. Governance
2. Business Impact Analysis
3. Risk Assessment
4. Recovery Strategies
5. Business Continuity/Disaster Recovery Planning
6. Test and Verification

Governance. Senior management involvement and support are critical to the success of a company's business continuity program. Executive buy-in enables the business continuity program to be in alignment with the company's strategic direction and business objectives. This also ensures that the program is able to obtain appropriate resources and visibility. Without adequate senior management involvement and support, a business continuity program risks losing effectiveness and alignment with business strategy, misspent or unfit resources, gaps between capability and requirements, or in the worst case, senior management eliminating business continuity altogether because they do not see the value in the investment.

A key component for governance is the creation and enforcement of business continuity standards and policies. These standards and policies outline the *what* and *how* of business continuity. This allows for program consistency across the company and supports corporate audits. The governance board has the responsibility to support and oversee the business continuity program.

No company can implement a robust business continuity program overnight; it can take years for a complex global company to fully implement a business continuity program. Business continuity is a journey that must be evaluated, maintained, and aligned with an organization's three- to five-year strategy. The governance board is responsible for business continuity oversight and direction; the board is in charge of the journey.

Business Impact Analysis (BIA). A BIA is a methodology to identify critical business processes and functions based on operational and/or financial impacts. This is accomplished by interviewing business process owners and asking them to describe their business processes. This interview includes the identification of critical resource requirements (staff, equipment, etc.), vital records and data, along with internal and external dependencies. Analysis of the data gathered through these interviews paints a picture of the critical paths within a business at any given time. This step also identifies the business threshold for disruption loss, including applications, systems, platforms, and infrastructure.

The business impact analysis identifies the preliminary recovery time objective (RTO)[3] and recovery point objective (RPO).[4] It is important to remember when designing a business continuity solution that it is not restoring business to normal, but it is the restoration of what is most crucial at a given time. For example, if the company issued payroll the day previous to an "event," restoring the payroll process would not be critical. But if payroll was to be released the day after the event then restoring the payroll process would be critical, especially to employees. The business process owners also describe work-around procedures that can be implemented until the process can be resumed or the staff can return to work.

Risk Assessment. The risk assessment stage identifies business continuity risks that could result in a business process disruption or hinder recovery. A risk assessment usually includes a facility assessment and an environmental analysis. A high-level physical inspection of a facility should include a review of the electrical design, mechanical heating ventilation and air-conditioning (HVAC) design, communications and network architecture review, physical security evaluation, emergency egress/ingress, and structural design of the data center and call center (as applicable). The environmental risk analysis includes the analysis of the likelihood of natural and man-made disasters at a specific location. After the risks are identified, they should be ranked and rated by criteria specified in the business continuity standards.

Recovery Strategies. The data gathered from the BIA and risk assessment portray the existing business continuity capabilities and gaps. Recovery strategies are developed to mitigate these potential risks. Recovery strategies and the associated estimated costs for implementation are developed and presented to the business continuity governance board for review. It is up to the governance board to approve and fund the chosen recovery strategies. Note the governance board should also sign off on high-ranked business risks with the reasoning on any decisions not to remediate a risk.

Business Continuity/Disaster Recovery Plans. Business continuity planning allows for the availability of critical business processes in the event of an incident that renders facilities, computer systems, and/or employees inoperable or inaccessible. The goal of the creation and implementation of business continuity and disaster recovery plans is to minimize economic losses resulting from disruptions to business functions. These plans provide steps and procedures to facilitate an orderly recovery of critical business functions and/or systems. Business continuity plans address the recovery of business functions and workspaces; disaster recovery plans address the recovery of the information technology environment and systems that support the business. The provisions in these types of plans are used as the basis for providing guidance, preparing for, and effecting recovery activities in connection with executive management's discretion. Tactically, the business continuity/disaster recovery plans address how to do the following:

- Minimize business losses resulting from disruptions to business processes
- Provide a plan of action to facilitate an orderly recovery of critical business processes and technical infrastructure
- Identify key individuals or teams who will manage the process of recovering and restoring the business and/or technology after an incident or disaster
- Specify the critical business and technical activities that need to continue after an incident
- Outline the logistics of recovering critical business processes and technical infrastructure

Proper execution of these plans facilitates the timely recovery of critical business processes. Business continuity and disaster recovery plans are effective only if they are maintained properly and the content information is current. A key element of business continuity/disaster recovery plans is the coordination between information technology and business processes to align RTO and RPO with business requirements over time.

Test and Verify. The business continuity standards will guide the business continuity program's roadmap to the development, testing, and maintenance of continuity and disaster plans and reporting to the governance board. The tests are used to train associates and create an awareness of the business continuity program model and individual roles. This is done through exercising the plan. Different levels of plan testing range from tabletop "walk-throughs" to the actual mobilization of plans. Actual mobilization of plans requires increased resources but will provide more thorough results.

The key to a thriving business continuity program is that it is never stagnant. It is a living process, and as it matures it should evolve into being part of regular business operations, not viewed as simply an add-on.

BCP Exercises

A tabletop or structured walk-through exercise is a paper evaluation of a portion of a business continuity plan without the expenses or personnel resources associated with a full test. The exercise scope can vary from a review of a portion of the BCP to a review of the entire plan. The walk-through has many worthwhile objectives:

- Verify the contents of the plan
- Prepare for simulation testing
- Train new members and create employee awareness
- Maintain preparedness while limiting use of resources
- Affirm that the strategy documented in the plan is viable
- Educate critical personnel on their responsibilities in a disaster
- Confirm that the information in the plan is current and accurate
- Identify areas of the plan that need revision or updates

The primary benefit of a tabletop exercise, besides offering the opportunity to realize an impressive set of objectives, is that it is cost-effective and noninvasive.

A second type of exercise, called a component exercise, is usually performed during off-hours and tests a particular segment of the recovery plan. It differs from the structured walk-through in that it involves actual recovery activities. The overwhelming benefit of a component exercise is that it is nondisruptive and focused. Various types of component tests can include the following:

- Tests of the emergency notification system
- Evacuation tests
- Data center or application recovery test
- Remote or dial-in access test
- Critical business function recovery test

A mobilization exercise is an integrated simulation/full operations test that includes an exercise performed at the actual recovery sites and utilizing backup resources that would be used during an actual event. A structured walk-through and/or a component exercise test should precede the mobilization exercise. The primary objective of a mobilization exercise is to test an entire plan or a portion of the plan under emergency scenarios, validate operational effectiveness and business unit interdependencies, and provide technical and administrative measurable results. Measures of test results should be compiled during the exercise and then compared against expected results.

An exercise of this proportion is normally scheduled to take place after hours or during a weekend. While the most costly in terms of resources, the major benefit of a mobilization exercise is that it requires interdepartment coordination and is the best true test of the business continuity program.

After the exercise type, identification of recovery priorities, objectives, timeline, and scenario have been determined, a company conducts the test, analyzes the findings, and develops corrective actions. The final step is to update the business continuity/disaster plan to incorporate lessons learned from testing.

CONCLUDING THOUGHTS

The supply chain management concept has come a long way from the days of materials management, inventory cycle counting, expediting orders, and fighting fires. The concept has become a profession and many outside the industry, including executive managers, have come to the realization that supply chain excellence is a critical success factor for business. However, many of the risks we have discussed within the arena of business operations are still traditionally mitigated and managed by reactionary metrics and methodologies. As supply chain risk management unfolds from a concept into a more mature discipline, many of the new tools, techniques, mitigation strategies, and metrics that we present will become an effective way to identify, assess, mitigate, and manage operational risks.

Summary of Key Points

- Operational risks are by far the most frequent number of risk events, not only for those who manufacture products, but for service organizations as well. Operational risks are contained in each of the four pillars of supply chain risk that include supply risk, demand risk, process risk, and environmental risk.
- Operational risks cross several planning and execution horizons. The two most prevalent horizons are the tactical horizon, which

normally covers 1–18 months into the future, and the operational horizon, covering 0–45 days into the future.

- Supply risk can be further classified into supplier risk, logistics risk, and fraud, corruption, and counterfeiting risk.
- Demand risk consists of customer risk, product risk, and logistics risk.
- The categories for the process risk pillar are known or hard risks, unknown or soft risks, and chronic risks that can arise within a company's four walls.
- The fourth risk pillar, environment/ecosystems risk, is probably the most immature pillar and also contains known, unknown, and chronic risk categories.
- One traditional and effective proactive approach to identifying, assessing, mitigating and managing operational risks, business continuity planning, has been around for many years. It is still considered a cost-effective approach to being prepared for and responsive to risk events, particularly operational risk events.
- The business continuity life cycle includes six stages: governance, business impact analysis, risk assessment, recovery strategies, business continuity/disaster recovery planning, and test and verification.
- A tabletop or structured walk-through exercise is a paper evaluation of a portion of a business continuity plan without the expenses or personnel resources associated with a full test. A second type of exercise, called a component exercise, is usually performed during off-hours, tests a particular segment of the recovery plan, and involves actual recovery activities.

ENDNOTES

1. Accessed from The European Banking Authority (EBA) and The European Bank for Reconstruction and Development (EBRD) at www.eba.europa.eu and www.EBRD.com.
2. Accessed from www.dictionary.com.
3. RTO = Recovery Time Objective: The maximum tolerable time to recover critical business functions and the existing resources that support each function.
4. RPO = Recovery Point Objective: The maximum amount of data loss allowable.
5. The authors would like to acknowledge the contribution of Betty Barnes to this section.

8

Supply Chain Fraud, Corruption, Counterfeiting, and Theft

Groeb Farms, a major U.S. honey supplier, was involved in the nation's largest food fraud: an incident that eventually forced the company into bankruptcy. This honey laundering case involved a German company, ALW, that illegally imported cheap and sometimes adulterated Chinese honey. While the honey itself is not illegal, mislabeling the product and avoiding tariffs is a different story. ALW had its offices in China use independent brokers to procure honey that eventually became honey "not from China."

The motivation behind the honey fraud is that honey from China to the United States is subject to tariffs that can triple its cost. Fifty-gallon drums of Chinese honey procured by ALW were shipped to India, Malaysia, Indonesia, South Korea, Russia, Mongolia, Thailand, Taiwan, and the Philippines, where they were relabeled and often filtered to remove traces of their origin. Fake documents would then be produced to move the honey through U.S. customs. Groeb Farms, based in Michigan, fired two executives who produced fake documents and lied to their board of directors about the origin of the honey the company was procuring, even when the company's internal auditors raised concerns that the honey was illegally imported. Groeb Farms paid a $2 million fine and was required to dispose of the Chinese honey it had in inventory—something that was very expensive considering that the replacement costs of honey were rising at the time. Amid the company's financial troubles, it fell behind on payments on a bank loan. U.S. honey producers and distributors who claimed that they were harmed by Groeb's actions filed several lawsuits also. This fraud, besides being ethically and legally wrong, clearly exposed the company to risk.

In this chapter we explore fraud, corruption, counterfeiting, and theft and discuss how these issues impact everyone from the supplier to the consumer. We'll provide examples of companies that have been accused and convicted of fraud, corruption, or counterfeiting and think about the financial impacts emanating from these risk events. We will present some new terminology and address the vast amount of new rules and regulations coming from many countries. And we'll share new tools, techniques, tactics, solutions, and methodologies to help stamp out corrupt practices.

SOME KEY CONCEPTS

A detailed report on fraud, corruption, and counterfeiting, which surveyed more than 114,000 professionals in 107 countries, came up with a startling finding. The study concluded that one in four people had recently paid some form of a bribe over a 12-month period.[1] While this seems shocking, we have to remember that not all countries or cultures view this topic the same way. Fraud, corruption, counterfeiting, and theft are a large part of supply chain risk.

The following provides a quick summary of some key concepts that are important parts of this chapter.

Bribery

Bribery is trying to persuade someone, typically illegally or dishonestly, to act in one's favor by offering a gift of money or other attractive inducement. While supply chain managers in the United States know that bribery is illegal, the same is not true around the world. In some countries, bribery is referred to as "facilitating payments," a term that is much more benign sounding. In Greek, the word *fakelaki*, literally meaning "little envelope," is used in Greek popular culture as a term referring to the bribery of public servants and private companies by Greek citizens in order to "expedite" service. Not that long ago the Greek parliament actually made this practice legal.

In many countries bribery is an accepted way of doing business. More than three out of four people in Liberia and Sierra Leone, for example, said they paid a bribe over a 12-month period. Bribery rates are more than 50% in Cambodia, Senegal, Cameroon, Ghana, India, Tanzania, Kenya, Libya,

Mozambique, Uganda, Yemen, and Zimbabwe. Conversely, Australia, Belgium, Portugal, Malaysia, Finland, Denmark, and Croatia have the lowest bribery rates at less than 5%. In the United States and UK this figure is just over 6%.

A study from Compliance Week and Kroll concluded that, on the positive side, large corporations based in the United States take anticorruption programs more seriously than their counterparts based elsewhere.[2] On the less-positive side, almost half of all respondents report they conduct no anticorruption training with their third parties. Of those who do train their third parties, only 30% believe their efforts are effective.

A headline on a prominent website stated that Ralph Lauren admitted to bribery at its Argentina subsidiary.[3] A Ralph Lauren subsidiary allegedly bribed customs officials to improperly obtain paperwork necessary for goods to clear customs and to avoid inspections. Also, the company was accused of faking invoices to mask payoffs. The U.S. Securities and Exchange Commission, in a nonprosecution agreement, fined the company $1.6 million. The fine would probably have been larger, but the company, according to the SEC's enforcement director, responded appropriately to the violations and cooperated fully with the investigation. The company's response was that bribes are inconsistent with the culture of compliance and integrity of the company. Still another corporate response was there was no evidence that the improper activity in Argentina was known or authorized by anyone outside of Argentina or that similar practices were occurring at other foreign operations.

Counterfeiting

A specific type of intellectual property theft involves counterfeiting. A counterfeit is something made in imitation so as to be passed off fraudulently or deceptively as genuine.[4] The sad truth is that many supply chains are plagued by counterfeits. Although movies, books, records, clothing, and other consumer goods are regularly copied, counterfeit goods are increasingly making their way into industrial supply chains. This includes products such as pharmaceutical drugs, automobile parts, and aerospace replacement parts that place the health and safety of individuals at risk.

Examples of counterfeiting globally are all too common. Car owners in the United States who have had their air bags replaced in the last several years, for example, run a high risk that the replacement bags are counterfeit. The U.S. Department of Defense says defense supply chains are

TABLE 8.1

Losses to Counterfeit Goods
by Country, U.S. Dollars

United States	$225 billion
Mexico	$75 billion
Japan	$75 billion
China	$60 billion
Germany	$32.25 billion
Canada	$30 billion

Source: www.havocscope.com.

inundated with counterfeit parts, and efforts to curtail them are failing. The U.S. Congress estimates there could be a million counterfeit parts in the supply chain. Besides the risk of product failure, Chinese counterfeit components could offer a backdoor to cyber snoops, escalating the threat of cyberspying and intellectual theft.[5]

One of most well-known and respected departments in the U.S. government, the Nuclear Regulatory Commission (NRC), is very concerned about counterfeit parts making their way into nuclear energy plants within the United States—so much so that the NRC is driving the issue of counterfeiting with its Counterfeit, Fraudulent, and Suspect Items (CFSI) project. We'll talk more about the NRC's projects and their countermeasures later in the chapter. For now, let's look at a few key data points on counterfeit parts relative to nuclear energy around the globe. Globally, a procurement director of a Russian supplier to the nuclear power industry was arrested for buying low-quality raw materials and pocketing the difference. During this same period multiple South Korean nuclear plants identified numerous counterfeit issues and shut down over fake control cable certification. The South Korean president ordered a parliamentary investigation.

Tables 8.1 and 8.2 present the estimated scope of global counterfeiting across various product categories and countries. As these tables reveal, the costs of counterfeiting to the U.S. economy, and to global industry as a whole, are significant.

Fraudulent, Corrupt, Coercive, and Collusive Practices

Fraudulent practices are any action or omission, including misrepresentation, that knowingly or recklessly misleads or attempts to mislead a party to obtain a financial benefit or to avoid an obligation.[6] *Corrupt practices* means the offering, giving, receiving, or soliciting, directly or indirectly,

TABLE 8.2

Counterfeit Goods by Industry, U.S. Dollars

Counterfeit drugs	$200 billion
Counterfeit electronics	$169 billion
Software piracy	$63 billion
Counterfeit foods	$49 billion
Counterfeit auto parts	$45 billion
Counterfeit toys	$34 billion
Music piracy	$12.5 billion
Counterfeit clothing	$12 billion
Counterfeit shoes	$12 billion
Cable piracy	$8.5 billion
Video game piracy	$8.1 billion
Counterfeit sporting goods	$6.5 billion
Counterfeit pesticides	$5.8 billion
Mobile entertainment piracy	$3.4 billion
Counterfeit cosmetics	$3.0 billion
Movie piracy	$2.5 billion
Counterfeit aircraft parts	$2 billion
Counterfeit weapons	$1.8 billion
Counterfeit watches	$1 billion
Fake diplomas and degrees	$1 billion
Total for all goods worldwide	$652 billion

Source: www.havocscope.com.

of anything of value to influence improperly the actions of another party. *Coercive practice* means impairing or harming or threatening to impair or harm, directly or indirectly, any party or the property of the party to influence improperly the actions of a party. And finally, *collusive practice* means an arrangement between two or more parties designed to achieve an improper purpose, including influencing improperly the actions of another party.

Let's look at some headlines coming out of China over the last several years. These announcements from China came through several U.S. media outlets:

- GlaxoSmithKline was placed under scrutiny for bribing doctors and hospital officials in efforts to sell products at higher prices; several executives were detained.
- Fines are being administered to six dairy companies in China accused of price fixing and anticompetition practices. The fines will amount to a record setting $108 million!

- Sanofi, France's largest pharmaceutical company, was accused of paying bribes totaling 1.69 million yuan to 503 doctors at 79 hospitals in Beijing, Shanghai, and Hangzhou.
- Novartis was placed under investigation for possible bribery and fraud in the provision of medical drugs and services. A Novartis employee claimed her manager instructed her to use 50,000 yuan (US$8,000) to boost sales of its drugs.

The standard response to these cases is one of initial shock and then full cooperation with the investigating authorities of the country where the risk event occurred. Over the years, insurance and other risk management organizations have been monitoring these events and others. And with so many observations, we've developed a Four-Stage Model of how most risk events play out, regardless of the event cause or country of origin, which Figure 8.1 illustrates.

As seen in this figure, the first stage of a risk event tends to be denial. Normally it means a belief the event is not that damaging and a company will be back in business in a few days. In 2013, the U.S. attorney general announced fines levied against Johnson & Johnson for false marketing of three drugs, Risperdal, Invega, and the heart failure drug Natrecor. The attorney general stated the company marketed the drugs for unapproved use and paid "kickbacks" to doctors and nursing homes. The company's response was typical in terms of damage control and our Four-Stage Model—J&J continued to stand by Risperdal as safe and effective for its approved use.

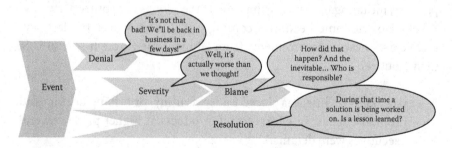

FIGURE 8.1
The four stages of a fraud, corruption, or supply chain disaster.

The second of the four stages, severity, talks about the real scope and scale and normally includes realization that the event is actually worse than first thought. The third stage, blame, is the stark reality that the event is a total mess and follows this line, "We can't understand how this happened and are working diligently to solve the problem." All the while, executives are asking questions such as, "How did this happen and who's responsible?" Finally, the fourth stage, resolution, should be a teaching moment in terms of supply chain risk management.

What we've witnessed, along with other research organizations, academia, and consulting firms, is that supply chain risk management and to some extent general risk management is still an *ad hoc* activity. And when the event has been resolved and the company is still in business, those who handled the event go back to their normal job responsibilities. Most company executives assume that their direct reports are prepared to handle risk events. This assumption, however, is flawed. If you fail to prepare and train for risk events, do not be surprised that the probability of a good outcome from a risk event is suspect.

RULES AND REGULATIONS

We would like to present the growing array of rules and regulations already in place and in the works that attempt to diminish, but not necessarily eliminate, illicit and fraudulent activity. Whatever your opinion is regarding rules and regulations, the demand to be in compliance is at its highest level ever and probably will not subside for the foreseeable future. Table 8.3 presents a sample of some of the newest and farthest-reaching rules and regulations impacting global supply chains. As you can see, there is no shortage of rules and regulations.

We've identified a few new rules, regulations, and organizations to drill down a bit deeper in an effort to provide additional perspective.

Consumer Financial Protection Bureau (CFPB)

The CFPB is a relatively new organization in the U.S. government enacted to ensure that products are not mislabeled or maliciously marketed to U.S. consumers and to protect consumers from fraud and price gouging.

TABLE 8.3

Rules and Regulations

Regulation/Law/ Organization	Definition
AEO	Authorized Equipment Operator
AES	Automated Export System
BIS	Bureau of Industry & Security
CBP	Customs & Border Protection
CFPB	Consumer Financial Protection Bureau
CSI	Container Security Initiative
C-TPAT	Customs Trade Partnership against Terrorism
Dodd-Frank Wall St Reform & Consumer Protection Act	Financial reform and conflict minerals impacts
EEI	Electronic Export Information
ETS	European Union Emissions Trading Scheme
FCPA	Foreign Corrupt Practices Act
FSMA	Food Safety Modernization Act
GSP	Generalized System of Preference
ISF	Import Security Filing
PIP	Partners in Protection
REACH	Registration, Evaluation, Authorization & Restriction of Chemical
RoHS	Restriction of Hazardous Substances Directive
Sarbanes-Oxley	Financial Reporting Transparency Act
WEEE	Waste Electrical & Electronic Equipment Directive

Customs Trade Partnership against Terrorism (C-TPAT)

This program is the U.S. Customs and Border Protection's (CBP's) premier trade security program. The purpose of C-TPAT is to partner with the trade community for the purpose of ensuring the U.S. and international supply chains are not subject to intrusion by terrorist organizations. C-TPAT requires the trade company participants to document and validate their supply chain security procedures in relation to existing CBP criteria or guidelines. CBP requires that C-TPAT companies develop an internal validation process to ensure the existence of security measures documented in their Supply Chain Security Profile and in any supplemental information provided to CBP. The purpose of the validation is to ensure that the C-TPAT participant's international supply chain security measures contained in the C-TPAT participant's security profile have been implemented and are being followed in accordance with established C-TPAT criteria and guidelines.

Dodd-Frank Wall Street Reform and Consumer Protection Act

The Dodd-Frank legislation was a direct result of the 2008–2009 financial meltdown. This act also includes subjecting banks to stress testing to evaluate their financial resiliency. It is a far-reaching and somewhat controversial law. The legislation also includes Conflict Minerals Act, which took effect in 2014.

In 2012 the SEC approved a final rule requiring companies to disclose their use of conflict minerals (tantalum, tin, tungsten, and gold) and whether those minerals originated in the Democratic Republic of Congo or adjoining countries.[7] These regulations will require companies to trace various materials to their source to ensure they did not originate in conflict regions. The compliance disclosure is required by the U.S. government. Companies are required to disclose that they verified the suppliers' address, audited those suppliers, and required those suppliers to certify that the materials incorporated into the manufacturers products comply with the laws. Customers will be asking manufacturers to certify their products through the global supply chain. We wish them luck.

Foreign Corrupt Practices Act

Under the Foreign Corrupt Practices Act (FCPA) it is unlawful for a U.S. firm, as well as any officer, director, employee, or agent of the company, to offer, pay, or promise to pay money or offer anything of value to any foreign official for the purpose of obtaining or retaining business. It is also unlawful to make payments to any person while knowing that all or a portion of the payment will be offered, given, or promised directly or indirectly to any foreign official for the purpose of assisting the firm in obtaining or retaining business. Firms are subject to fines of up to $2 million, while officers, directors, employees, agents, and stockholders are subject to fines up to $100,000. Under federal criminal laws other than FCPA, individuals can be fined up to $250,000 or up to twice the amount of the gross gain or loss of a transaction. And finally, a person or firm found in violation of FCPA may be barred from doing business with the federal government.

TOOLS, BEST-IN-CLASS PRACTICES, AND COUNTERMEASURES

Our focus in this section is to provide some insights on emerging tools that are providing capabilities to manage critical aspects of supply chain risk. This section explains the four quadrants presented in Figures 8.2 and 8.3.

Fraud, Corruption, and Theft Tools

The left portion of Figure 8.2 relates to a solution called Decision Point from Navigant. This tool is Navigant's antibribery onboarding portal. It is designed for the investigation of distribution and supply chain third parties. The tool addresses the challenges inherent in utilizing intermediaries and suppliers, such as onboarding, regulatory compliance and education, information gathering, approvals, and investigative processes as they relate to antibribery and other compliance regulations. The tool facilitates the third-party onboarding process, assesses and risk scores third party corruption risk, and enables due diligence investigations when appropriate in an effort to support the U.S. FCPA, Foreign Corrupt Practices Act.

The right portion of Figure 8.2 provides a workflow of a supplier onboarding process, which Navigant and other antibribery and corruption tools maintain for clients. As you can see, the tactical tasks are interwoven into a company's overall supplier management process. These tasks inside Navigant's Decision Point and other tools are co-managed processes rather than fully outsourced. For example, supplier onboarding may be initiated and managed by a commodity manager or buyer in the procurement group, who then sends the invitation, interacts with the supplier, and monitors progress. These tasks could then trigger key tasks at each step done by a third party, such as launching a survey, requesting certificates, validating responses, and more.

Moving to the left side of Figure 8.3 we profile another emerging solution tool from IHS that captures significant data for manufacturers across the globe in terms of where a supplier is located, its market share, inventory metrics, financial disposition, and financial risk. Looking at IHS's tool kit, you might consider it a supply chain "temperature check."

The right side of Figure 8.3 highlights a new solution set from Verisk Crime Analytics, in conjunction with C. H. Robinson, which monitors

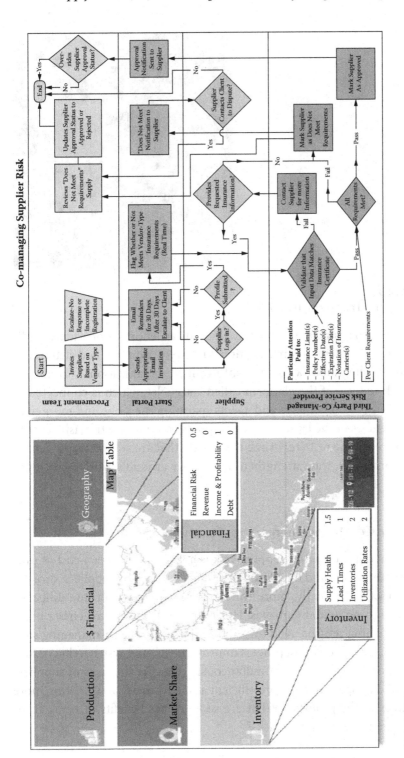

FIGURE 8.2
Emerging risk management tools.

FIGURE 8.3
Emerging fraud and corruption tools.

and alerts companies around the globe in an effort to provide coordinated theft incident communications, recovery support, and deterrence measures integrated with law enforcement organizations worldwide. Through the use of the platform called CargoNet, which is connected to more than 120,000 officers and agents spanning the globe and more than 9,000 agencies in the United States, shippers can mitigate the risk of cargo theft by receiving near-real-time theft alerts that are automatically sent to law enforcement, thus strengthening security around sensitive and high-value freight.

Supplier Co-Management

We know that managing supplier risk is a trade-off between cost of diligence and the cost and likelihood of damages caused by suppliers, directly or indirectly. Normally, more time and money spent on due diligence results in lower actual damages. However, over time this reaches a point of diminishing returns. And therein lies the key to the co-managing supplier concept. Disciplines, process improvements, and systems that reduce the cost of diligence while simultaneously reducing the risks of supplier-induced damages are far more efficient and effective than throwing more resources at the problem. What are some of the benefits of using third parties to manage supplier risks?

- **Economies of Scale**—Third parties that serve hundreds of companies can realize efficiencies not possible by individual companies, and therefore data can be collected once and used many times.
- **Economies of Specialization**—These tasks are core competencies for third parties and therefore they become efficient and effective over time at performing them.
- **Investment in Technology**—Again, because these are core competencies, third parties continue to invest heavily in platforms and systems to streamline the processes and provide overwhelming value.
- **Accumulated Expertise**—Experienced providers have learned what works and have created templates and libraries that can lead, guide, direct, and coach their clients.
- **Regulatory Currency**—Often providers maintain a cadre of experts to keep up with regulatory changes, denied party lists, and legislation to ensure their clients stay in compliance.
- **Improved Initial Risk Awareness/Identification**—More risks are exposed in the due diligence processes, and therefore action can be taken quickly to avoid risk.
- **Increased Number of Suppliers Managed**—The disciplines, processes, and systems brought by an experienced third party can generate huge improvements in the number and percentage of suppliers for which the proper due diligence is performed in depth, thus reducing incidents and damages.
- **Enhanced Early-Warning Risk Alerts**—Finally, with well-developed monitoring tools, emergent risks can be discovered sooner, giving the client more time to deal with an event and execute an optimal response.

Addressing Corruption with Best Practices

The U.S. Conference Board and the Center for Responsible Enterprise and Trade executed an exhaustive global survey involving Fortune 500 companies such as DuPont, Microsoft, ARAMARK, Emerson, Caterpillar and more.[8] This study found that 70% of respondents felt there was an extensive risk of corrupt activities when working with agents and business partners in emerging countries. Furthermore, an agreed-upon effective compliance program should consist of seven pillars: risk assessment, due diligence, contract provisions, audit and monitoring, governance and management, codes and policies, and training and communications. The study also found that training of company employees is a key ingredient for compliance success. More than 60% of respondents felt their company would benefit greatly from third-party, independent rating of their anti-corruption programs.

This report also found that the emerging market where companies utilize suppliers and agents the most is China. Other top markets for suppliers and agents are India, Southeast Asia, Mexico/South America, and Eastern Europe. If your supplier base is scattered across these top five countries/regions, which is probably the case, we know there's enough risk to go around for everyone. How do companies address corrupt practices with suppliers?

- **Perform Due Diligence**—The methods utilized the most by respondents include examining legal and financial aspects, reviewing of government relations, criminal investigations as a risk assessment technique, reviewing policies of third parties, and third-party ethics reviews.
- **Contract Provisioning**—Verbiage and actions most often placed in contracts include complying with laws and regulations, right to terminate by the company, right to terminate if third party violates laws, third-party financial liabilities, right to access supplier records, complying with anticorruption policies, and the right to audit the supplier.
- **Audit/Monitor**—Methodologies and processes most often in effect are searches for criminal activity, review of payments with high-risk suppliers, reviewing policies of third parties, periodic supplier and third-party recertification, and periodic audits and visits.
- **Rely on Policies**—Policies most often in effect are anticorruption information for employees, training of employees, reporting for employees, ongoing compliance programs, and reports on and for third-party employees.
- **Rely on Procedures**—Procedures most often in effect are constant communication about corruption, written policies for third parties, multiple approvals for third parties, company employment agreements, bids reviewed by company teams, and anticorruption information for third-party companies.

- **Governance and Responsibilities**—In this arena, the internal audit, compliance, finance, and legal departments are extensively involved in anticorruption efforts. Often, responsibility for anticorruption efforts with suppliers and third parties will reside with the procurement group.

Counterfeit Countermeasures

Counterfeiting, as mentioned earlier, is a topic of interest across some important industries. It's a major issue, especially to the Nuclear Regulatory Commission (NRC). The NRC is one of the biggest proponents of supply chain risk management inside the U.S. government. When the NRC speaks, everyone listens, including other countries. Below are some key methodologies the NRC advocates as well as tools, techniques, and countermeasures to support its ongoing battle against counterfeit parts.

The NRC advocates the diligent use of procurement policies and procedures, as we witnessed in the Conference Board's report. It also stresses the issue of parts, manufacturers, and distribution facilities having audits and assessments. The NRC has developed for its nuclear clients an obsolescence product life cycle protocol covering components from cradle to grave. The commission has also developed several measures for counterfeit part risk detection and new traceability verification methods using serialization and tags. And finally, the NRC has worked to bolster law enforcement resources and penalties.

The NRC has developed countermeasures to detect counterfeit items. It has introduced new techniques such as Fourier transform infrared spectroscopy (FTIR), thermo-mechanical analysis, and X-ray fluorescence spectroscopy (XFR). Something that is particularly leading-edge is SigNature DNA.

SigNature DNA is something completely different in terms of counterfeit countermeasures. It is a plant-based electronic signature marking that is invisible to the naked eye. The markings show up only under ultraviolet light. Each supplier marks its product with its own customized sequence of which the combinations and permutations are infinite. Thus, the Pentagon is paying extra for these suppliers to mark all electronic components during a trial period. Also, the SigNature DNA technique is now expanding into new industries such as wine and high-end apparel.

Another new approach to mitigate the risk of counterfeit electronic components is growing rapidly. According to a report by the U.S. Senate Armed Services Committee, titled "Inquiry into Counterfeit Electronic Parts in the Department of Defense Supply Chain," an overwhelming majority of more than a million counterfeit parts identified in the investigation were sourced from a party other than the original manufacturer or an authorized distributor.

In the electronics industry, and now others as well, after a major risk event or a problem with a key supplier within an industry, buyers often have no choice other than to begin sourcing in what is called the gray market. A gray market is an unauthorized channel of distribution. However, turning to the gray market and purchasing a part from a source other than the original manufacturer or authorized distributors exposes the buyer to not only substandard components but also increases their risk of purchasing counterfeit components. As a result we have seen the birth of the excess and obsolete (E&O) industry. Franchised E&O distributors with guaranteed product traceability are filling a gap in supply chains resulting from just-in-time manufacturing and supply disruptions. By holding excess and unsold factory stock and guaranteeing product traceability back to the manufacturer, these distributors offer buyers a credible and reliable source of product to mitigate counterfeit risk.

While the approaches presented here are not the only ways to counter the counterfeiters, they illustrate some of the creative thinking that is occurring to try to stay one step ahead in an area that is a continuous concern.

CONCLUDING THOUGHTS

This chapter highlights the darker side of human nature. When we think about natural disasters, we think about things called "acts of God." These acts are not caused directly by humans, although climate change proponents will argue that this is not necessarily the case today. Something we do know for certain is that every time an act of fraud, corruption, counterfeiting, or theft occurs, it is a man-made event. And man-made events are preventable.

The problem with fraud, corruption, counterfeiting, and theft, particularly at a national level, is that these behaviors, particularly corruption, correlate highly with poorer and less-successful countries. As a predictor of failure, at one level corruption may have the most devastating effect of any risk presented in this book. As the Austrian journalist Karl Krause once commented, "Corruption is worse than prostitution. The latter might endanger the morals of an individual. The former invariably endangers the morals of the entire country."[9]

Summary of Key Points

- Bribery is trying to persuade someone, typically illegally or dishonestly, to act in one's favor by offering a gift of money or other attractive inducement.
- A counterfeit is something made in imitation so as to be passed off fraudulently or deceptively as genuine.
- The first stage of a risk event tends to be denial. The second of the four stages is severity (scope and scale). The third stage is blame (reality sets in). The final stage is resolution (where the lesson is learned).
- New organizations enacted to cut down on fraudulent activities include Consumer Financial Protection Bureau and Customs Trade Partnership against Terrorism. New regulations with this same goal include the Dodd-Frank Wall Street Reform and Consumer Protection Act and the Foreign Corrupt Practices Act.
- Tools against fraud and corruption include Navigant's antibribery onboarding portal called Decision Point, IHS's "temperature check," and Verisk Crime Analytics' CargoNet.
- An effective compliance program should consist of seven pillars: risk assessment, due diligence, contract provisions, audit and monitoring, governance and management, codes and policies, and training and communications.

ENDNOTES

1. Riley, Charles. "Governments Lose Trust As Corruption Soars." *CNNMoney*, July 9, 2013, Accessed from www.cnnmoney.com.
2. *Compliance Week Magazine* and Kroll Anti-Bribery & Corruption Report, May 2013 Survey.
3. O'Toole, James. "Ralph Lauren Admits Bribery at Argentina Subsidiary." *CNNMoney*, April 22, 2013, Accessed from www.cnnmoney.com.
4. Accessed from www.dictionary.com.

5. Fulghum, David, Bill Sweetman, and Jill Dimascio. "China Chips: Counterfeit Components Reveal Political Hype and Bureaucratic Muddle in Washington." *Aviation Week and Space Technology*, June 4–11, 2012: 68.
6. European Banking Board terms and definitions.
7. Accessed from http://www.pwc.com/us/en/cfodirect/publications/dataline/2012-10-sec-adopts-conflict-minerals-rule-public-and-nonpublic-companies-in-many-industries-are-affected.jhtml.
8. USA Conference Board & Center for Responsible Enterprise & Trade Report, August 2012.
9. Accessed from http://www.brainyquote.com/quotes/quotes/k/karlkraus152098.html.

9

Emerging Risk Management Frameworks for Success

Our focus in this chapter will be on emerging frameworks that are being leveraged to drive successful supply chain risk management (SCRM) initiatives. We will become grounded with basic definitions and explore some of the new frameworks, standards, and rules and regulations that frame the supply chain risk management landscape. We'll then profile the frameworks from several research organizations' perspectives and present several leading companies who are utilizing these frameworks to implement risk initiatives within their organizations. We'll conclude by highlighting several benefits to be derived from utilizing these frameworks.

WHAT IS A FRAMEWORK?

A framework is a skeletal, openwork, or structural frame. This term also describes a frame of reference, which includes an arbitrary set of axes with reference to which the position or motion of something is described or physical laws are formulated.[1] One professional organization profiles the term *framework* in several perspectives. One perspective provides a concept revolving around organizational design by viewing a framework as an organizational structure to support the strategic business plans and goals of an enterprise (e.g., for-profit and not-for-profit companies). Given the mission and business strategy, the organizational structure design provides the framework within which operational and management activities will be performed. A second perspective revolves around the operating environment and views a framework as the global, domestic, environmental, and stakeholder influences that affect the key competitive factors, customer needs, culture, and philosophy of each individual company. This environment becomes the framework in which business strategy is developed and implemented.[2]

FRAMEWORKS SUPPORTING THE NEW SUPPLY CHAIN RISK MANAGEMENT DISCIPLINE

Whether you are in operations, finance, distribution, banking, or academia, several frameworks are critical for supply chain risk management. Recall that Chapter 1 defined SCRM, which is expanded here to refer to the implementation of strategies to manage everyday and exceptional risks within the supply chain through continuous risk identification, assessment, mitigation, and management with the objective of reducing vulnerability and ensuring sustainability. We view SCRM as the intersection of supply chain management and risk management. Let's discuss several of the critical frameworks.

Enterprise Risk Management (ERM) Framework

As mentioned in Chapter 1, the general ERM framework has been around for many years, emanating from the finance and classical risk insurance disciplines. We'll take a high-level view at ERM first, and then dig deeper with profiles from CAS, the Casualty Actuarial Society. Recall that Chapter 1 provided one perspective of ERM. A second perspective is from CAS, which has defined ERM as the discipline by which an organization in any industry assesses, controls, exploits, finances, and monitors risks from all sources for the purpose of increasing the organization's short- and long-term value to its shareholders.

ERM can also be described as a risk-based approach to managing an enterprise, integrating concepts of strategic planning, operations management, and internal control. ERM is still evolving to address the needs of various stakeholders who want to understand the broad spectrum of risks facing complex organizations and their supply chains to ensure they are appropriately managed. Government regulators and debt-rating agencies have increased their scrutiny of the risk management processes of many companies.

COSO ERM Framework

An important perspective about risk is put forth by the Committee of Sponsoring Organizations of the Treadway Commission (COSO), a well-known group formed to help businesses develop their internal control systems. Thousands of organizations have incorporated COSO's Internal Control Integrated Framework to help manage their activities. In 2001, in response to a heightened awareness of global risk, COSO partnered with PriceWaterhouseCoopers to develop a framework that would enable organizations to evaluate and improve enterprise risk management. COSO defines ERM as follows:

> A process, effected by an entity's board of directors, management and other personnel, applied in a strategy setting and across the enterprise, designed to identify potential events that may affect the entity, and manage risk to be within its risk appetite, to provide reasonable assurance regarding the achievement of entity objectives.[3]

Eight interrelated components comprise COSO's ERM framework. These components are derived from the way management runs an enterprise and are integrated within the management process. These eight components, which are also relevant to our discussion of SCRM, comprise a fully developed ERM system:

- **Internal Environment.** The internal environment sets an organization's tone, including how risk is viewed and addressed by an organization's people, including its risk management philosophy, risk appetite, integrity, and ethical values.
- **Objective Setting.** Enterprise risk management ensures that management has a process to set objectives and that the chosen objectives support the entity's mission and are consistent with its risk appetite.
- **Event Identification.** Internal and external events affecting the achievement of objectives must be identified, distinguishing between risks and opportunities. Opportunities are channeled back to management's strategy or objective-setting processes.
- **Risk Assessment.** Risks are analyzed in terms of their likelihood and impact. This is used as a basis for determining how to manage risks.
- **Risk Response.** Management selects various risk responses, including avoiding, accepting, reducing, preventing, or sharing risk. A set

of actions are developed that align risks with the entity's risk tolerances and risk appetite.

- **Control Activities.** Policies and procedures are established to help ensure risk responses are carried out.
- **Information and Communication.** Relevant information is identified and communicated in a form and time frame that enable people to carry out their responsibilities. Effective communication flows down, across, and up the organization.
- **Monitoring.** The entirety of enterprise risk management is monitored and modifications are made as necessary. Enterprise risk management monitoring is accomplished through ongoing management activities, separate evaluations, or both. Management makes modifications to the ERM plan as required.

ISO Standards

Most of us probably know something about the International Organization for Standardization (ISO) standard organization, but for those of you who are not familiar with this standards body, we'll start with some basic foundational elements of this worldwide organization. Founded in 1947 in Geneva, Switzerland, ISO is an international standard-setting body composed of representatives from various national standards organizations to promote worldwide proprietary, industrial, and commercial standards. The official languages of the ISO are English, French, and Russian. The organization adopted the abbreviation ISO based on the Greek work *isos* (meaning equal) as its universal short form name of their organization.

The organization known today as ISO began in 1926 as the International Federation of the National Standardizing Associations (ISA), whose focus was mainly on mechanical engineering. It was disbanded in 1942 during World War II but was reorganized under its current name in 1947. ISO is a voluntary organization comprising 163 member countries, whose members are recognized authorities on standards, each one representing one country. The bulk of the work of ISO is done by 2,700 technical committees, subcommittees, and working groups. Each committee and subcommittee is headed by a secretariat from one of the member countries. ISO is funded by a combination of (1) organizations that manage specific projects or loan experts who participate in technical work, (2) subscriptions from member bodies, which are in proportion to each country's gross national product, and (3) the sale of the standards' work products. With that as our backdrop regarding the organization, let's talk about what this standards-setting body has developed relative to our SCRM discipline.

ISO 31000. The purpose of this standard, introduced in 2009, is to provide principles and generic guidelines on risk management. It seeks to provide a universally recognized paradigm for practitioners and companies employing risk management processes to replace the myriad of existing standards, methodologies, and paradigms that differed between industries, subject matters, and regions. The scope and intent of this standard is to provide generic guidelines for the design, implementation, and maintenance of a risk management process throughout any organization, regardless of industry. The standard is designed to enable all strategic, management, and operational tasks of an organization, through projects, functions, and processes, to be aligned to a common set of risk management objectives.

The implementation of this standard is to be applied within existing management systems to formalize and improve risk management processes as opposed to wholesale substitution of legacy management practices. When implementing ISO 31000, attention should be given to integrating existing risk management processes into the new paradigm addressed in the standard. The focus should be centered around the following:

- Transferring accountability gaps in ERM
- Aligning objectives of the governance frameworks with ISO 31000
- Embedding management system reporting mechanisms
- Creating uniform risk criteria and evaluation metrics

Using ISO 31000 can help organizations increase the likelihood of achieving their objectives, improve the identification of opportunities and threats, and effectively allocate and use resources for risk management. ISO 31000 cannot be used for certification purposes but does provide guidance for internal or external audit programs. Organizations can compare their risk management practices against internationally recognized benchmarks for effective management and corporate governance.

A Risk Insurance and Management Society (RIMS) survey of risk professionals found that 22% of firms use the COSO standard as their ERM framework, while 23% follow the international ISO 31000 standard.

Twenty-six percent of firms say they do not follow a particular standard or framework.[4] A large percentage is not sure or has nothing significant in place.

ISO 73. This new Risk Management Vocabulary standard, updated in 2009, provides a wide breadth of terms. This standards body has been updating the vocabulary recently to take into account the growing need for additional terms and taxonomy within global supply chains. Some commonly used risk terms in this standard are risk management, risk assessment, risk analysis, risk, risk source, risk evaluation, risk criteria, risk avoidance, risk transfer, risk reduction, risk mitigation, risk retention, risk optimization, risk acceptance, risk financing, risk control, risk communication, risk perception, stakeholder, and interested party, just to name few. Many of these terms have been defined in our earlier chapters and will be discussed in subsequent chapters as well.

Besides the ISO standard, the new Supply Chain Council supply chain risk model, residing in the new SCOR 11.0, is available. The SCOR community has performed a comprehensive update to its supply chain models, metrics, and terminologies, including an updated view of supply chain risk.[5] APICS has also aggressively developed a body of knowledge covering SCRM for members and customers.

ISO 28000. This standard is also new. It was developed in 2010 and is actually a series of standards, all under the umbrella of 28000, which broadly covers the requirements for a security management system within the supply chain. The standards inside 28000 are 28001, 28002, 28003, 28004, and 28005. You may not have stumbled into this standard as of yet because it's actually listed under "Ships and Marine Technology" on the ISO website. This is not surprising to us, because most of today's global trade is done by cargo ships circling the globe in a complex pattern. Nonetheless, the ISO 28000 series of standards are applicable to all modes of transport, air cargo included, considering all the threats within that industry and others. We'll briefly introduce you to all the standards in this series and then profile 28002 individually.

- 28001—Best practices for implementing supply chain security, assessments and plans, and requirements and guidance
- 28002—Development of resilience in the supply chain
- 28003—Requirements for bodies providing audit and certification of supply chain security management systems

- 28004—Guidelines for the implementation of ISO 28000
- 28005—Electronic Port Clearance (EPC) part 1 and part 2

Published in September 2010, ISO 28002 covers security management systems for the supply chain and the development of resilience in the supply chain. Resilience is the adaptive capacity of an organization in a complex and changing environment. It also describes the capability of an organization to prevent or resist being affected by an event or the ability to return to an acceptable level of performance in an acceptable period of time after being affected by an event. This newly published standard attempts to provide insights into how an organization can engage in a comprehensive and systematic process of prevention, protection, preparedness, mitigation, response, continuity, and recovery.

Jan Husdal, an early and prolific SCRM blogger, has done follow-up work on these ISO standards and has provided various process maps, which provide us a perspective on how the standards group is looking at both internal and external supply chain security.[6] Figure 9.1 is an illustration of one such map for ISO 28002. Husdal notes that the process maps are similar to the SCOR model approach.

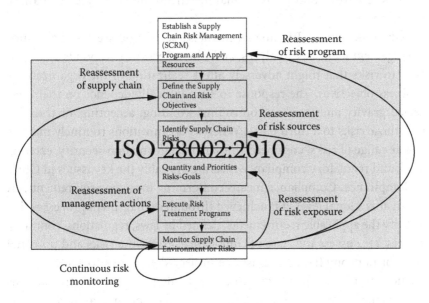

FIGURE 9.1

ISO 28002. (Source: Husdal, Jan, SCRM Blog, 2013. http://www.husdal.com/2010/11/04/iso-28002-supply-chain-resilience/.)

Governance, Risk, and Compliance (GRC)

The GRC framework has been around for some time. Through discovery, this framework has been continuously scrutinized and criticized as somewhat ill-defined. However, much more rigor has been spent recently reviewing and solidly codifying this framework. The next segment attempts to provide some context on this subject, which we feel supports the foundation for successful SCRM. The following describes the three basic tenets of this framework: governance, risk management, and compliance.

Governance. Governance describes the overall management approach through which senior executives direct and control the entire organization, using a combination of management information and hierarchical management control structures. Governance activities ensure that critical management information reaching the executive team is sufficiently complete, accurate, and timely to enable appropriate management decision making and provide the control mechanism to ensure that strategies, directives, and instructions from management are carried out systematically and effectively.[7] Aberdeen Group has synthesized this definition by saying that governance includes the frameworks and tools, policies, procedures, controls, and decision-making hierarchy employed to manage the business.[8]

Risk Management. Risk management is a set of processes through which management identifies, analyzes, and where necessary responds appropriately to risks that might adversely affect realization of the organization's business objectives. The response to risks typically depends on their perceived gravity and involves controlling, avoiding, accepting, or transferring those risks to a third party. Whereas organizations routinely manage a wide range of risks, commercial/financial, information security, external legal, and regulatory compliance risks are arguably the key issues in GRC.

Compliance. Compliance means conforming to stated requirements. At an organizational level, it is achieved through management processes that identify the applicable requirements, defined by laws, regulations, contracts, polices, etc.; assess the state of compliance; assess the risks and potential costs of noncompliance against the projected expenses to achieve compliance; and hence prioritize, fund, and initiate any corrective actions deemed necessary. Aberdeen Group views compliance as meeting the required or mandated regulations that are governmental, industry specific, or internally imposed.[9]

With much more focus on risk, many research organizations, such as Aberdeen Group, AMR (now Gartner), and others have revisited the GRC framework. It seems apparent to some that executives are viewing effective compliance and risk management as opportunities for corporate growth, keeping in mind that customers and partners will always choose to do business with a company possessing fewer liabilities. Furthermore, being aggressive in building a business is about taking risks, so by having an effective risk management structure in place, a company can essentially be bolder in addressing new market opportunities. And finally, compliance is crucial in establishing new grounds for business, such as global or regional expansion, which requires companies to meet a strict set of guidelines in order for the company to conduct successful business. The following quote sums up well the importance of the GRC framework:

> The challenges with risk management are in embedding an understanding of the risk management process, ownership of risks within the business, and the cultural change required for a truly risk-aware decision-making culture rather than being seen as a compliance obligation. To overcome these challenges we have been conducting risk management training for all staff, increasing engagement and constantly iterating in all communications that risk management is to assist the business in achieving objectives.
>
> **Risk and Compliance Manager**
> *Liberty International Underwriter*

A set of primary objectives underlie those companies that are best-in-class in terms of utilizing the GRC framework. These companies:

- Drive the organizational alignment of executive and staff agendas through effective governance
- Understand risks in terms of dollar-value impact and corporate brand equity
- Prioritize organizational initiatives based on risk type and risk level of severity
- Create additional revenue opportunities by meeting compliance requirements for selling into new markets/regions

A set of strategic capabilities needed to achieve bottom-line results from a GRC framework include promoting accountability within the organization

Pressures	Actions	Capabilities	Enablers
• Increase in regulatory requirements	• Promote accountability within the organization through effective communication • Provide visibility and access to dynamic regulatory requirements	• Standardized workflow for risk identification and mitigation • Systematic monitoring of key risk indicators • Centralized repository for risk information & data • Standardized procedure to communicate management direction	• Governance, risk & compliance solutions • Risk management tools • Workflow automation • ERP, Enterprise Resource Planning • Safety compliance solutions • Environmental solutions • Financial modeling • IT security solutions • Regulatory portals • Sustainability solutions • Supply chain management • EPM, Enterprise Performance Management

FIGURE 9.2
Best-in-class GRC framework.

through effective communications, providing visibility and access to dynamic regulatory requirements, standardizing work flow for risk identification and mitigation, systematically monitoring key risk indicators, and centralizing risk information and data. Figure 9.2 illustrates Aberdeen's profile of a best-in-class GRC framework. Table 9.1 provides some new performance measures emerging within the GRC framework.

TABLE 9.1

Governance, Risk, and Compliance Metrics

GRC Metric	GRC Measurable Values
Year-over-year change in risk value	Percentage change in risk value in the past 2 years (*risk value* is defined as monetary equivalent of the liability)
Year-over-year change in compliance-related cost	Percentage change in compliance-related cost in the past 2 years (e.g., cost of delayed production, recalls, stop-shipments, fines, penalties incurred from non-compliance)
New market revenue	New-market revenue, as a result of compliance, as a percentage of total revenue in the past 12 months
Compliance audit success rate	Percentage of compliance audits that yielded positive results in the past 12 months
Governance effectiveness	Percentage of management directives executed successfully in the past 12 months

Source: Aberdeen Group, "Effective GRC Management: Strategies for Mitigating Risks and Sustaining Growth in a Tough Economy," May 2012.

We will end our GRC conversation with some comments from a senior risk manager at McKesson, the nation's oldest and largest health care services company. The senior manager of IT governance, risk, and compliance at McKesson provides his view about the GRC framework when he says that GRC is about organizational collaboration including internal audit, technology risk management, compliance groups, legal, and more. He further argues that most companies are faced with organizational and functional silos, poor integration, lack of visibility, wasted resources, unnecessary complexity, and wasted information.

Over the past few years, McKesson has acquired a number of companies. Each acquisition has required McKesson to take on a new set of challenges in terms of developing an integrated platform. McKesson's risk manager maintains it is difficult to reduce cost if you don't have an integrated view of the activities within your organization, something the GRC framework demands. This means sometimes you have to step away from the tactical tools and process controls. If your leaders are not visionary and don't understand what they don't know, this risk manager argues you have serious challenges ahead. The visionary leadership at McKesson has enabled the risk management team to make great strides toward an integrated GRC platform. This senior risk manager argues a company must have visionary leadership, communications, an enterprise-wide perspective, fact-driven analytics, and stakeholder engagement to be successful. If a company maintains these basic GRC elements, the end result will be unprecedented transparency and visibility, the ability to make risk-based decisions, accountability, and alignment across the business.[10]

RISK TAXONOMIES—AN OPERATIONAL FRAMEWORK FOR SCRM

We've mentioned several strategic frameworks that are critical success factors to an effective supply chain risk management discipline. To make managing an enterprise-wide risk management process simple and practical, we need to take complex material, break it down, and make it accessible to everyone in an organization. What is needed is the ability to build a more operationally oriented methodology, something we will refer to as a *risk taxonomy*. Taxonomy is the practice and science of naming, classifying, and defining relationships between resources, risks, goals, and business processes across an enterprise. Without risk taxonomies or a way to structure and classify risk events, it is difficult to understand different types of risks across the enterprise. And without taxonomies there is no common set of standards or way to manage relationships between different data types. If each area of the business uses its own terms to classify risk, then the aggregated information will be subjective, incomplete, redundant, or at best, flawed. Each silo in an organization and level within each silo will speak a different dialect.

The basic approach when creating a risk taxonomy is to develop a common framework for all risks, their readiness standards, and a balanced scorecard of objectives. To handle the complexity of a large-scale supply chain, this approach obviously requires a tool to effectively manage built-in libraries for use across the enterprise and highlight how one risk event in one functional area affects other functions. These tools enable the organization to create structured, centralized repositories of all risk elements within the organization. Some of these elements are risks, goals, requirements, relationships (vendors, customers, third parties), software applications, physical assets (buildings, servers, data centers, plants, equipments, and tools), data repositories, people, policies, and user-defined applications (models and spreadsheets). For each of these elements, taxonomy tools and techniques allow for flexibility and customization to manage cross-functional cause-and-effect relationships. Some basic capabilities of these taxonomy tools include the following:

- **Creating and Maintaining a Central Repository of Information**— This could include the use of predefined fields or completely customized data elements needed by the organization.
- **Full Document Management**—This should provide the ability to upload documents, link them to shared applications, with a version control aspect and permission rights so that all information related to these areas can be centrally stored.
- **Enterprise-wide Task Management**—From a more tactical perspective, this could provide for creating automatic reminder e-mail triggers for due dates, contract renewal dates, monitoring dates, approvals, and change notifications.
- **Risk Assessment Scoring**—In this area, tools can provide best-practice assessment factors or allow organizations to develop their own risk factors. With this capability, organizations can rate these elements to

determine priorities and criticality. They normally allow the company to also enter explanations for each of the assessments, thereby codifying the point-in-time assessment for future analysis and trends.

A risk taxonomy manages all the risk elements and links them to other elements within the organization to create a network of terms, definitions, and resource relationships. It codifies all the things that an organization should worry about before surprises occur, manages those things in one place with connections to provide assurance that these elements are actually being done effectively to mitigate risk. And in some cases, taxonomy tools provide the content to alert the organization to important changes within an industry and to be in a position to identify who and what resources are connected to or impacted by an industry or compliance issue.

LEVERAGING ERM, GRC, AND RISK TAXONOMIES

The importance of SCRM can't be stressed enough, as Ericsson found out in March of 2000. During this period, Ericsson, a leading mobile phone manufacturer, experienced a disruption in supply from Phillips Electronics. A lightning strike caused a fire at a Phillips facility in Arizona, resulting in the loss of millions of microchips and rendering this supplier dormant. Ericsson's production was totally disrupted because Phillips was the buyer's sole supplier of microchips. This disruption resulted in $400 million of lost sales and eventually caused Ericsson to exit the phone business. Conversely, Nokia, Ericsson's main competitor, had a multisource supplier strategy and quickly ramped up the production of microchips from another supplier. Nokia managed the supply chain risk and actually turned this risk event into an opportunity. After this risk event Ericsson implemented a risk management process that includes the identification, assessment, treatment, and monitoring of risks across its supply chain. The company created a corporate function called *corporate risk management* that consists of a council of members in supply and sourcing as well as members from each business area. Ericsson also created a risk management evaluation tool, which appears in Figure 9.3. This process looks at all areas of the supply chain, both internally and externally, along with

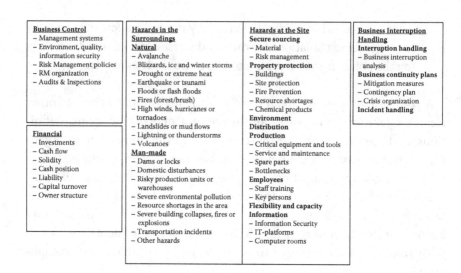

FIGURE 9.3
Ericsson risk management and evalulation tool (ERMET).

contingency planning to analyze risk exposure. Ericsson and Nokia are now two of the most ardent advocates of SCRM and actually don't talk much about their integrated SCRM approaches because they both consider these tools, techniques, and methodologies a strategic advantage.[11]

Leggett & Platt, Inc., a 125-year-old manufacturer of sleep technology that introduced the first bedspring and now designs and produces a diverse array of products for homes, offices, and vehicles, took a risk more than 10 years ago and introduced an ERM project across its entire organization.[12] In the mid-1990s, a company vice-president attended several ERM classes facilitated by the RIMS organization and felt the ERM process would benefit Leggett & Platt. However, the concept languished until the CFO raised the topic of implementing an ERM program. The company quickly formed a committee to launch the program.

The ERM committee consists of the functional heads at the corporate level, including the CFO, treasurer, and vice-presidents of IT, tax, legal, audit, and accounting. Each functional head identified internal and external risks in their own disciplines. They then assessed those risks in terms of severity and frequency. The committee continuously categorizes these risks, tracks them, plots them, and reports on them at every committee meeting. The committee now rates all risks and correlates them against other risks and operational key performance indicators (KPIs). Some lessons learned include (1) risk is a big part of business and if you don't take risks, you limit your potential for success; (2) taking on too much risk threatens a company's survival; (3) categorize risks in terms of severity, develop treatments for those different risk issues, and overtly manage those risks; and (4) without an ERM framework, a company does not have a process that is predictable and sustainable to identify, assess, mitigate, and manage risk.

From a GRC perspective, one company that stands out is Bayer Crop Science. Led by the director of forecasting and Sales and Operations Planning (S&OP), the company has developed a comprehensive approach for managing risk throughout its global supply chain. The forum used by Bayer Crop Science is its S&OP process. The framework they use is the classic GRC framework supported by the SCOR model.[13] According to the director of forecasting, risk management plays an integral part in the execution of Bayer's S&OP process. This approach allows the business to get a better feel for potential dangers and the impact they may have on the business. Bayer Crop Science is also an advocate of the GRC framework presented earlier in the chapter.

Another company focusing on SCRM and exercising diligence in terms of developing and maintaining a risk taxonomy is Coca-Cola. The formal SCRM group at Coca-Cola is driven by three directors of supply chain risk. Having an actual corporate group structured to drive supply chain risk and led by SCRM directors is still novel. The SCRM group utilizes many of the SCOR model elements, which include many of the Supply Chain Council's risk protocols, process maps, and metrics. The key aspect of Coca-Cola's approach to SCRM is its dedication to classifying and categorizing all risks within the company's global supply chain. Coca-Cola classifies and categorizes risks based on severity, treating risks differently, and maintaining a strict methodology to classify its risks. How do they do this? The company has built what it calls "risk registers." Every business unit maintains its own risk register, every region maintains a rolled-up or aggregated risk register, and every risk register is rated and compared with a corporate risk tolerance table before action is taken. The risk registers are updated and reviewed quarterly by the SCRM group. From a 50,000-foot level, Coca-Cola classifies risks primarily into strategic and operational risks, which Figure 9.4 illustrates.

The actual risk register identification and assessment process operates as follows. When a risk event occurs, employees access the online, worldwide

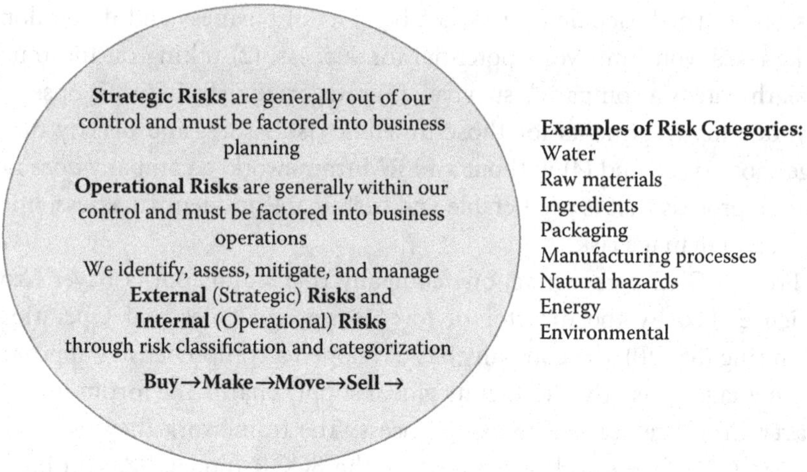

Strategic Risks are generally out of our control and must be factored into business planning

Operational Risks are generally within our control and must be factored into business operations

We identify, assess, mitigate, and manage **External** (Strategic) **Risks** and **Internal** (Operational) **Risks** through risk classification and categorization

Buy→Make→Move→Sell →

Examples of Risk Categories:
Water
Raw materials
Ingredients
Packaging
Manufacturing processes
Natural hazards
Energy
Environmental

FIGURE 9.4

Risk classification at Coca-Cola. Source: MIT/Coca-Cola presentation by Dr. Bruce Arntzen, director Global SCALE Risk Initiative—MIT, and John J. Brown, director risk management—Coca-Cola, "Current and Future State of Corporate Supply Chain Risk Management," *Supply Chain World North America*, May 25, 2011.

risk register system to first evaluate if their business unit or region has ever dealt with this type of risk before. If so, they immediately review all the pertinent information stored in the system in terms of how the business unit or region "treated" that risk and how long it took to mitigate the risk. If the unit or region has never encountered the risk, they search the world-wide risk register system to see if another unit has encountered this risk. If the corporation has never encountered the risk, a call to the SCRM corporate group is made, and collectively the teams begin the mitigation and management process. Without a diligent approach to risk taxonomy, the organization would not be able to quickly and effectively mitigate risks across the enterprise and around the globe.

BENEFITS OF ERM AND GRC FRAMEWORKS

An exciting benefit of utilizing ERM as an SCRM framework comes from AON (a leading global provider of risk management, insurance and reinsurance brokerage, human resources, and outsourcing services) and the

Wharton School of the University of Pennsylvania. Using annual financial results and Bloomberg market data for 361 publicly traded companies, these researchers found a statistical link between higher levels of risk maturity and higher relative stock price returns along with lower levels of stock price volatility and higher relative levels of return on equity performance.[14] The companies rated highest in maturity exhibited +18% stock return performance as opposed to the lowest rated companies, who demonstrated a negative stock return of –10%. A second performance indicator was return on equity.

Companies with the highest risk rating exhibited a return on equity of +37%, while organizations with the lowest rating produced a negative return of –11%. This differential between best and worst is the most dramatic metric in the study. And the researchers didn't stop there. They took the financial data and subjected that data to "stress resting" by simulating how securities would respond in the immediate aftermath of significant risk events to the financial markets based on historical data.

The researchers essentially conducted "shock therapy" on the data for companies in the study by modeling the Japanese earthquake and tsunami in 2011. Organizations with the highest maturity rating exhibited a stock price return of –0.3% over a certain period compared with organizations with the lowest rating exhibiting a return of –3.4%. We feel this speaks volumes for why companies should spend time and resources on ERM and other risk management frameworks. Although risk management can be a hard sell, these numbers are convincing when it comes to a solid SCRM ROI.

An additional study published by RIMS asked 564 organizations to participate in an in-depth assessment of ERM. The participants compared their ERM activities against a comprehensive set of best practices and readiness indicators inside a risk maturity model. The premise of the study was the belief that better-managed companies tend to have higher credit ratings and higher ERM competency. Credit ratings for participating companies were compared using statistical analysis to measure the relationship between credit rating scores and risk maturity model scores. The correlation coefficient was calculated for each model factor and found to be positive. The researchers also conducted statistical analyses that compared the model scores of two groups, those using ERM and those not using ERM. The researchers found statistical differences between the two groups that supported the value of ERM. Overall, the researchers' report concluded the following:

- Organizations with formalized ERM programs have higher risk maturity model scores (as we would expect).
- Organizations with higher risk model scores have higher credit ratings.
- Organizations without formalized ERM programs have lower risk maturity model scores.
- Organizations without formalized ERM programs have lower credit ratings.

Additional benefits from utilizing ERM as a risk framework from the RIMS study include the following:

- Companies can avoid potential future rating agency downgrades and increased cost of capital since Standard & Poor's and many other rating agencies have incorporated ERM into their business models.
- Companies can minimize the personal liability of board members and risk of criminal charges against executives for failure to act responsibly in making Sarbanes–Oxley quarterly certifications against fraud.
- Companies can meet regulators' expectations leveraging ERM and in turn minimize incremental compliance costs that can negatively impact the bottom line.

Finally, we'd like to share some relevant statistics on benefits derived from utilizing GRC as a risk framework from the Aberdeen study referenced earlier in the chapter.[15] This study reveals that top-performing companies that leverage the GRC framework experienced a 34% reduction in risk value and a 23% reduction in compliance-related costs over a two-year period. Those who lag in the use of GRC are much more likely than best-in-class companies to lose money on compliance investment, while best-in-class GRC companies are much more likely to obtain a positive ROI from their compliance initiatives. And best-in-class GRC companies are 54% more likely than their competitors to systematically evaluate business processes for compliance and 29% more likely than their competitors to conduct quantified risk assessments. These are compelling statistics.

CONCLUDING THOUGHTS

Part of the reason for discussing risk frameworks and taxonomies is to illustrate the evolving nature of SCRM into a bona fide business discipline. How else can we tell that SCRM is evolving into a legitimate discipline? Research organizations are developing supply chain risk frameworks and taxonomies and supporting these with research and metrics of success; academic organizations are starting to teach the concepts and providing additional research; and standards organizations are codifying standards around terms, definitions, processes, protocols, and measures of success. Furthermore, large consulting firms are writing white papers on supply chain risk management as Fortune 500 companies are executing those concepts to mitigate and even prevent supply chain risk. When this all occurs each and every day, it is safe to conclude that this thing called supply chain risk management just might be the real deal.

Summary of Key Points

- Frameworks provide a frame of reference for disciplines to operate successfully, whether in operations, finance, distribution, banking, or academia.
- ERM is a management framework that is critical to the success of SCRM. It can be leveraged to support the identification, assessment, mitigation, and management of strategic, tactical, and operational risks.
- GRC is another framework being embraced by many organizations to support SCRM initiatives. This framework should be considered an overarching approach to managing enterprise risk.
- The ISO organization and standards have been around since the 1940s. It's encouraging when a standards organization, made up of professionals from around the globe, begins to embrace a concept such as SCRM with new standards for terminology, best practices, security, and resiliency.
- A risk taxonomy is the practice and science of naming, classifying, and defining relationships between resources, risks, goals, and business processes within an enterprise. Without risk taxonomies or a risk breakdown structure or operational risk event classification, it is difficult to compare different types of risks across the enterprise. This critical, yet sometimes neglected, success factor to SCRM provides a

common set of standards or a methodology to manage relationships between different types of data and risks.

• Bottom-line benefits, including hard and soft ROIs demonstrate dramatically why organizations embrace risk frameworks to ensure a successful risk management journey.

ENDNOTES

1. Accessed from Webster's Dictionary.
2. Accessed from APICS Dictionary.
3. Accessed from *Enterprise Risk Management—Integrated Framework*. 2004. http://www.coso.org/documents/coso_erm_executivesummary.pdf.
4. Teach, Edward. "The Upside of ERM." *CFO*, November 2013: 44.
5. Accessed from SCOR, The Supply Chain Council, https://supply-chain.org.
6. Accessed from Husdal SCRM Blog, http://www.husdal.com/2010/11/04/iso-28002-supply-chain-resilience/, 2013.
7. Lamm, Blount. "Under Control: Governance across the Enterprise." Accessed from http://www.amazon.com/2013.
8. Aberdeen Group. "Effective GRC Management: Strategies for Mitigating Risks and Sustaining Growth in a Tough Economy Report." May 2012.
9. Aberdeen Group, May 2012.
10. As cited in Aberdeen Group, May 2012.
11. Daniels, Yanika, and Timothy Kenny. May 2008. "Leveraging Risk Management in the Sales & Operations Planning Process." Submitted for MS of Engineering in Logistics, Massachusetts Institute of Technology Engineering School, Certified by Dr. Larry Lapide.
12. Sleeping Better with ERM. *RIMS Magazine*, 60, 7 (September 2013): 18-9.
13. Brewer, Curtis, Director of Forecasting for Bayer Crop Sciences. *"Injecting Risk Management into the S&OP Process."* IBF Conference, 2011.
14. Accessed from AON Risk Maturity Index Insight Report, November 2013.

10

Using Probabilistic Models to Understand Risk

AMR research, now part of Gartner, has been speaking about the complexion of the 21st century supply chain for some time, and during that dialogue the topic of probabilistic planning continuously arises. This planning process is supported by stochastic demand management and dynamic inventory planning.

In this chapter, we will discuss models that have been around for some time, such as stochastic/probabilistic models, deterministic methods, discrete-event simulation, and digital modeling. We'll also explore how these methods are being leveraged to map out complex supply chains and how leaders are appending risk assessments to scenarios supported by these techniques. Next, we'll introduce risk response planning, the logical outcome of stress testing complex supply chains and modeling "what-if" scenarios in an effort to develop a plan to manage risk scenarios. We conclude with several examples that demonstrate how leading companies are leveraging these powerful and dynamic techniques to identify, assess, mitigate, and manage supply chain risks.

DEFINING THE MODELS

Stochastic/probabilistic models are models where uncertainty is explicitly considered in the analysis. Furthermore, stochastic/probabilistic models are procedures that represent the uncertainty of demand by a set of possible outcomes (i.e., a probability distribution) and that suggest inventory management strategies under probabilistic demands.[1] Stochastic

TABLE 10.1

Defining Key Terms

Technique	Description
Time series analysis	Deterministic approaches that use historical data to forecast future requirements.
Regression analysis	Deterministic models that represent the relationship between a dependent variable [y] and independent variables [x].
Discrete-event simulation	DES models the operation of a system as a discrete sequence of events in time. Each event occurs at a particular instant in time and marks a change of state in the system. While simulations allow experimentation without exposure to risk, they are only as good as their underlying assumptions.
Forecast error	Represents the difference between an actual value and a forecasted value. The objective is to minimize forecast error and maximize forecast reliability.
Stochastic/ probabilistic models	Models where uncertainty is explicitly considered in the analysis. Involves statistical procedures that represent the uncertainty of demand by a set of possible outcomes and that suggest inventory management strategies under probabilistic demands.
Design of experiments	The process of setting up a series of tests or experiments to determine what outputs result from different combinations of inputs.
Sensitivity analysis	Involves systematically changing quantitative inputs or assumptions to assess their effect on a final outcome.
Linear programming	A mathematical technique used in computer modeling (simulation) to find the optimal solution that maximizes profit or minimizes cost considering a set of limited resources, such as personnel, funds, materials, etc.

optimization (SO) methods are optimization algorithms that incorporate probabilistic (random) elements, either in the problem data (the objective function, the constraints, etc.) or in the algorithm itself (through random parameter values, random choices, etc.), or in both. This concept contrasts dramatically with the deterministic optimization methods, such as time series analysis, linear programming, integer programming, the simplex method, and regression models where the values of the objective function are assumed to be exact and the computation is completely determined by the values developed in the equations. (Table 10.1 provides basic definitions of some of the key terms used in this chapter.)

The most critical difference between probabilistic and deterministic models is that there is not an ounce of uncertainty explained or handled in deterministic tools. Therefore, the responsibility of handling any uncertainty, complexity, and risk has been the responsibility of supply chain professionals. At this point in the growth of supply chain management as a discipline and because of the expanded nature of uncertainty, complexity, and risk, it is important to use these "new" techniques to manage global risk. These methodologies are not new. Academia, pharmaceuticals, medical, finance, insurance, and the banking industry have been using these methods to evaluate and mitigate risk for more than 50 years. They are, however, new to the supply chain world.

With the framework established for stochastic/probabilistic methods, let's spend a brief moment describing our supply chain comfort zone in terms of tools.[2] Supply chain professionals have been leveraging deterministic methods to solve supply chain problems for more than 35 years. We utilize time series tools to forecast sales, basically using the least squares method to fit a line through a set of sales or order observations and project anywhere from 1 to 18 or more months of future sales. We regularly track forecast error by comparing actual demand versus projected demand and handle that demand variability (i.e., risk) with inventory safety stock, buffer stock, and simple brute force. We also utilize linear programming and the simplex method to optimize supply chain network designs by modeling existing and future network configurations and then optimizing an objective function to either maximize sales, profits or service levels, or minimize costs, subject to a series of constraints.

We also utilize, albeit sparingly, regression analyses to build models of our markets and attempt to predict sales for new product introductions, which are the dependent Y variables subject to independent X variables. And we've leveraged these tools to optimize revenues or minimize the costs associated with logistics, truck and rail scheduling, and airline operations management. While these approaches have merit, none handles uncertainty and risk. And in today's volatile world, that in itself is a compelling reason to act.

PROBABILISTIC VERSUS DETERMINISTIC MODELING TOOLS

This is a good point in our discussion to illustrate the differences between the two statistical methodologies and then follow up with some actual cases showing how probabilistic methods support the effort to manage risk within complex global supply chains. Think of this in terms of

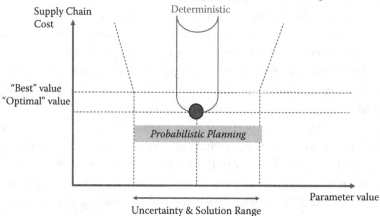

FIGURE 10.1
Stochastic/probabilistic planning.

weather forecasters on TV. When hurricane forecasters talk about a new storm, they present something they call the "cone of uncertainty," which Figure 10. 1 depicts.

This cone represents a set of outcomes from probabilistic models that attempt to predict where a storm will travel based on probabilities of occurrences. Compare this approach to the traditional deterministic methods where there is no uncertainty within the model. The bottom of Figure 10.1 depicts the extremely V-shaped solution that deterministic methods attempt to achieve, without uncertainty, in order to present an optimal solution. The probabilistic method, used by weather forecasters, provides an approach to gain an optimal solution across a much broader set of solution variables within the model, explicitly addressing uncertainty.

We are beginning to witness this probabilistic methodology supporting scenario planning in the context of supply chain risk management (SCRM). What does this process look like? First, it starts by digitizing the entire supply chain and building a flow model of the enterprise, as illustrated in Figure 10.2. Supply chains are nothing more than a network with speeds and feeds, inputs, outputs, and processing times that can be digitized as dynamic flow models. Next, companies populate the model of the enterprise with base case data from their enterprise resource planning (ERP) system and by identifying the historical behavior and uncertainty of all relevant factors. These factors include elements such as lead times, capacities, demand, production, inventory, quality, yields, policies, and more.

Companies next develop "what-if" scenarios, or what we call hypotheses, that need to be reinforced or refuted, looking at scenarios such as demand increasing by 30%, demand decreasing by 30%, lead times increasing or decreasing, market share to be gained, supplier disruptions, plant disruptions, complex competitive pricing changes, geopolitical changes, oil price fluctuations, and more. Most probabilistic tools maintain a library of probability tables that indicate the probability distributions utilized within the scenarios. If the tools can't capture historical data for certain variables, users can posit probabilities of occurrence for certain changes.

With these assumptions codified and historical data in hand, users begin to run discrete-event simulations across the entire enterprise for every scenario in an effort to review the cause-and-effect outcomes and their statistical strengths. The outcomes normally take the shape of histograms, sensitivity curves with confidence intervals and probabilities of occurrence along with risk assessments for each scenario, depicted on the bottom of Figure 10.2. This continuous running of the model, requiring several hundred iterations, can continue until the outcomes, per scenario, are considered statistically significant. This task is accomplished through the use of sensitivity analysis, optimized response curves, and design of experiments (i.e., a structured and systematic Six Sigma–oriented testing methodology of the process model).

The outcomes of the scenarios are next prioritized based on their probabilities of occurrence and their associated risk index. This novel approach is accelerating SCRM. By combining powerful tools, such as

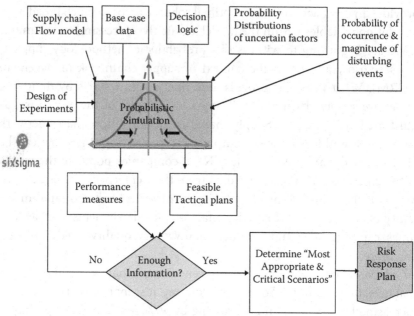

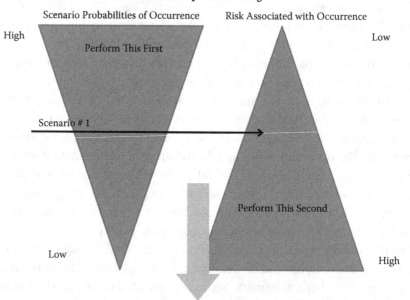

Take Scenarios & Build A Risk Response Plan (Perform This Last)

FIGURE 10.2

Scenario and risk response planning.

probabilistic methods, digital modeling, and discrete-event simulation, coupled with risk assessments for every scenario, these techniques are providing professionals the ability to better manage risk. The final step in this powerful new process is to eventually develop a risk response plan for the scenarios deemed critical to the enterprise covering the strategic, tactical, or operational horizons. This approach represents SCRM at its sophisticated best.

RISK RESPONSE PLANS

Industries such as oil and gas, chemicals, airlines, and pharmaceuticals and organizations such as the U.S. Department of Defense have been exercising scenario planning and supporting those scenarios with risk response plans for years. Formally, a risk response plan is a document defining the known risks including the description, cause, likelihood, costs, and proposed responses to risk events. It also identifies the current status of each risk.[3] Risk response plans tend to be the outcome of scenario planning exercises. As organizations develop scenarios to either stress test their supply chains in terms of exploring what-if cause-and-effect relationships or go through exhaustive scenarios to prepare for possible risk events, risk response plans are the value-added outcome of most of these risk management exercises. Risk response plans, as noted in Chapter 9, are one of the eight elements of the enterprise risk management (ERM) framework.

Some might argue that business continuity plans and risk response plans are similar. One key difference is that risk response plans tend to be outcomes of scenario planning exercises. Risk response plans are documents that profile basically what will be done if a certain risk event occurs, be it operational, tactical, or strategic. These plans normally consist of four areas:

- **Identification of Known Risks**. This involves a description about the nature of risks (refer to the ERM framework), the risk causes, the likelihood of risk occurrence as defined by the probability distribution and discrete-event simulation, and the estimated cost of each risk.
- **Identification of Risk Owners**. This involves identifying what disciplines and who from those disciplines, including existing roles and responsibilities, will assume leadership as risk owners.

- **Articulation of Risk Responses**. This part includes the response plans, what everyone will do in the event of a risk event, and what tactics to deploy. This might also include cost/benefit relationships.
- **Articulation of the Measure of Successful Risk Mitigation**. This includes the key performance indicators that reveal how well an organization is succeeding in mitigating a risk.

As discussed earlier, organizations that exercise scenario planning tend to force rank these scenarios using risk assessment techniques, which Chapters 12 and 13 will touch upon. They then determine the scenarios for which they will develop risk response plans. These risk response plans are then distributed, either in printed form or digitized, for distribution throughout a facility, division, or enterprise. This allows widespread access if a risk event occurs, something that should accelerate a recovery.

COMPANY EXAMPLES OF PROBABILISTIC MODELING

The following examples highlight the efforts of three leading companies to model risk probabilities into their risk management plans.

Scenario Planning at DuPont

DuPont, founded in 1802, has repeatedly transformed itself over its 200+ year history due to a culture of innovation and renewal. Today, DuPont is at the forefront of supply and demand chain management as it develops a comprehensive approach to sales and operations planning (S&OP). As mentioned in Chapter 3, S&OP is an essential process for balancing supply and demand. DuPont has been pursuing S&OP across many divisions starting after 2000. With that demand/supply balancing approach the company has been

- Relating improved demand management outcomes to business performance as the driver for improved supply chains
- Breaking down complexity into actionable parts
- Addressing uncertainty with consensus planning
- Clearly defining and developing standard practices, knowledge, and resources

- Developing a global deployment methodology including diagnostics and a current-to-future-state improvement path
- Measuring for results and enabling the process with systems and tools

Even with its risk management and planning capabilities, the financial meltdown of 2008 caught DuPont off guard. One of the reasons the company was negatively impacted, even though it operated a comprehensive, although somewhat traditional, S&OP process, was the company's inward- rather than outward-looking focus when engaging in supply chain planning. This inside-out approach proved to be inadequate for responding to the dramatic market shifts that took place during the financial crisis. This experience has made DuPont one of the biggest proponents of something called scenario planning within the S&OP process.

Scenario planning uses probabilistic methods to evaluate and plan for demand events over various time horizons. DuPont is actively developing demand management scenarios with best, conservative, and most likely outcomes, all supported with probabilities of occurrence. To say that DuPont has elevated the sophistication of its risk planning game would be a tremendous understatement. Figure 10.3 summarizes the DuPont approach.

Within its S&OP process, DuPont models demand projections with probabilities and develops risk response plans for these demand scenarios. Compared with an earlier era, many of these scenarios are "outside-in," meaning they actively incorporate external events and factors that may impact DuPont's supply chains. To the risk purists, this is called *stress testing* the supply chain. This approach is similar to the apparel industry, which uses probabilistic demand planning because of the nature of its seasonal demand requirements.

DuPont routinely uses the scenario planning approach and an outside-in view of its supply chains to mitigate risk within its mature S&OP process. A scenario planning approach to demand management is now embedded across many of the company's supply chains and has fostered a concerted effort to elevate the importance of SCRM. Developing a functioning S&OP process is a good thing. Developing an S&OP process that includes demand projections with probabilities and risk response plans is even better. According to a DuPont executive, "The value of this type of approach to demand and supply balancing and risk management, especially when linked to automation and facilitation, is that it helps create planning scenarios that are actionable and executable, not just academic exercises."[4]

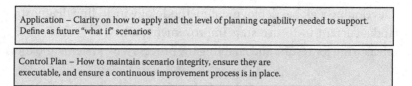

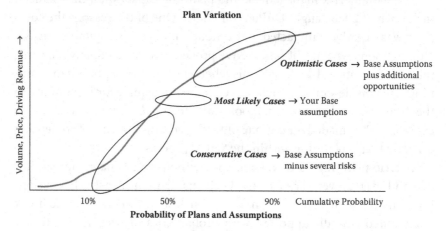

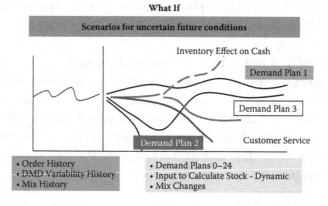

FIGURE 10.3
DuPont's approach to defining how to bring scenario planning to an actionable level.

Stress Testing the Supply Chain at Bayer Material Sciences

Another application we would like to discuss is stress testing complex supply chains utilizing the tools and techniques described in this chapter. The story we'd like to share is one revolving around a large and complex supply chain at Bayer Material Science. BMS is among the world's largest polymer companies. Its business is focused on manufacturing high-tech polymer materials and the development of innovative product solutions for its customers. BMS maintains more than 30 production sites around the globe.

BMS was encountering unacceptable levels of aged and slow-moving inventory across many of its business lines, and its compounding business wanted to significantly reduce its response times to customers in order to gain market share in regions they had never sold to in the past. Several hypotheses were established to better understand how the company's supply chains would react if these strategies were implemented. BMS established a Six Sigma team of black belts and operational analysts to quantify these hypotheses. The team developed a project to test these hypotheses under the DMAIC (define, measure, analyze, improve and control) umbrella. The DMAIC methodology is the most frequently used framework for evaluating and improving existing processes.

The BMS team realized early on that the company's complex manufacturing facilities would make it highly unlikely that the team would be able to test its hypotheses in live operations. Thus, attempting to execute trial production runs in live operations would be prohibitively expensive, time-consuming, and pose unacceptable risks to the organization. With that reality, the BMS team reached out to a third party to leverage its probabilistic/predictive modeling capabilities to test more than 40 hypotheses. Probabilistic/predictive digital models supported by discrete-event simulation and Six Sigma frameworks such as DMAIC, design of experiments, and failure modes effect analysis (FMEA) create a robust set of approaches for qualifying and quantifying the impacts of changing production points, policies, and procedures inside complex manufacturing environments.

The BMS team and the third-party consultants developed a digital model of the company's supply chain. The bulk of the data came from the company's ERP and supply chain systems. The company did a thorough job of developing its hypotheses and defining the existing environment. This was accomplished through the use of the Six Sigma methodology FMEA. The Bayer team also exercised lean/Six Sigma methods such as kaizen brainstorming events and value-stream mapping.

When utilizing FMEA the team developed a comprehensive tableau of rows and columns depicting more than 100 independent variables that the team felt were impacting customer service and aging inventory levels. These independent variables included elements such as the planning and scheduling process, sales forecasting process, partial orders, customer order errors, customer cancellations, overproduction, and missed commitments.

Each row of the FMEA tableau depicted a cause-and-effect relationship between an independent variable and the primary dependent variable, aged inventory. The team then established a prioritized activity impact profile of all variables based on three years of experience. The initial data were then entered into a probabilistic model in conjunction with historical orders, production, inventory, shipments, quality yields, and more. A digital model of the supply chain was tested for completeness through the use of postanalysis. This statistical validation method involves running hundreds of daily production schedules from actual data and comparing what the model predicted would happen with what actually happened. This resulted in a baseline that confirmed that the digital model was indicative of how Bayer operated its plants over a fixed period of time.

With the *define* and *measure* stages of the DMAIC process complete, the teams entered the *analyze* phase. In this phase the team captured the company's physical process, operating policies, and decision rules and incorporated them into the model. The team correlated operating performance with historical records, evaluated and compared alternative process improvement scenarios, and leveraged the probabilistic model to evaluate more than 40 scenarios. The team also incorporated the Six Sigma technique design of experiments, which they wrapped around the digital model to ensure statistically significant cause-and-effect relationships between the variables driving the outcomes of each scenario.

Many of the outcomes of the modeling runs were revealing, compelling, and at times even counterintuitive. The outcomes of the scenarios revealed that BMS could reduce lead times by a substantial amount to capture new market share; capture a significant amount of new market share without requiring new production capacity; increase their service levels; improve capacity throughput without new capital expenditures; and reduce aged inventory substantially while reducing working inventory as well.

Stress testing a complex supply chain before a company commits tremendous resources is a testimonial to the new tools coming online to support increased uncertainty, complexity, and risk in global supply chains. Leveraging digital modeling, probabilistic tools, discrete-event simulation, and risk assessment is a powerful environment to evaluate operational alternatives without experimenting on customers.[5]

Next-Generation S&OP at Huntsman

A final example involves Huntsman and its inability to react to sources of uncertainty that were impacting its S&OP process. So, what did the company do? Huntsman recruited two external consultants to help identify, assess, mitigate, and manage the sources of uncertainty that were negatively impacting the company's supply chain.[6]

Huntsman is a global manufacturer of differentiated chemicals. The company maintains an extensive array of production sites in Europe and the United States. With a complex supply chain and a product portfolio that runs the gamut of commodity chemicals with almost no margins to specialty chemicals with strong margins, its S&OP process to balance demand and supply was problematic and always involved managing difficult trade-offs. With a challenging operating environment to contend with, Huntsman came to the conclusion that something had to be done to manage the overwhelming number of sources of uncertainty in its supply chain if the company planned to compete effectively from a global perspective.

Let's get grounded a bit and talk about the company's supply chain and S&OP process. At one point Huntsman had just about every deterministic software enabler a company could imagine to plan, source, make, deliver, and return goods within a complex global supply chain.[7] Huntsman also maintained a sophisticated ERP system to take orders, make orders, ship orders, and bill orders. Within this IT environment, the company also managed a state-of-the-art demand management/forecasting tool with all the bells and whistles, including single, double, and triple exponential smoothing forecasting tools. Even today who wouldn't want these systems? These tools and systems acted as an enabler for a large-scale S&OP process covering a good portion of Huntsman's vast product portfolio.

Huntsman was also an avid user of lean/Six Sigma tools such as value-stream mapping, DMAIC, FMEA, and statistical process control charting. And the company maintained several new advanced planning systems (APSs) that took demand signals from the S&OP process and input them directly into the APS tools to create work plans across their worldwide manufacturing and distribution sites. They also maintained a huge data warehouse that enabled the company to slice and dice supply chain data into an almost infinite number of ways to support improved decision-making capabilities.

Most observers would agree the company was operating at a sophisticated supply chain level. Unfortunately, these deterministic approaches did not represent the next generation of risk management methodologies. Most of us are comfortable with what we know, and that has included the continued use of deterministic tools and techniques, none of which addresses risk. And therein was the dilemma for this large and sophisticated manufacturer. The deterministic models can only take us so far.

Huntsman was clearly looking to think outside the supply chain box as it recruited outside experts to help the company take a risk-oriented view of the supply chain and its processes in an effort to manage risk and improve bottom-line results. Let's look at how this change project began.

First, the two consultants developed several hypotheses, gathered data, and tested hypotheses. They attempted to identify the possible sources of uncertainty that were impacting Huntsman's S&OP process. Huntsman, as a lean/Six Sigma house, was already codifying many of the sources of uncertainties through the use of FMEA. The company was capturing all the known sources and then preparing frequency charts to profile these risks and provide insight into their cause-and-effect relationships. The consultants identified 13 sources of uncertainty impacting Huntsman's supply chain and its S&OP process.

A value-added activity from this project was taking the sources of uncertainty and classifying them into categories of low, medium, and high risk and then applying these sources of uncertainty across the three supply chain planning horizons. These time horizons include operational (0–45 days), tactical (1–18 months), and strategic (1–5 years). Figure 10.4 reveals what risks line up where and which time horizon maintains the highest number of risks. Figure 10.4 reveals that a large number of risks reside with the S&OP horizon (1–18 months). One external assessment has revealed that more than 70% of companies that practice S&OP remain in Stage I (reacting) and Stage II (anticipating) within an S&OP maturity model. Only 17% achieve Stage III (collaboration) status and a lesser amount achieve Stage IV (orchestrating). Clearly, there are opportunities for advancement here.

So, what did Huntsman do with this new-found insight on supply chain risks? Essentially, Huntsman and the project team relied upon many of the tools and techniques discussed earlier in the chapter, particularly probabilistic planning, digital modeling, discrete-event simulation, and risk assessment. The company tested hypotheses to determine the

Sources of Uncertainty	Operational 0–45 Days	Tactical 1–18 Months	Strategic 1–5 Years
Exchange rates	XXX	XX	
Supplier lead-times	X	XXX	X
Supplier quality	XX	X	
Manufacturing yield	XX	XX	
Transportation times	XX	XX	X
Stochastic costs	X	XXX	XX
Political environment			XX
Customs regulations	X	XX	XXX
Available capacity	XX	XX	X
Subcontractor availability	XXX	XX	
Information delays	XXX	XX	
Stochastic demand	X	XXX	XX
Price fluctuations	X	XXX	X

X – Low; XX – Medium; XXX – High

FIGURE 10.4
Sources of uncertainty in the supply chain.

cause-and-effect relationships of many of the codified sources of uncertainty and applied risk assessments to each scenario. With the outcomes of the modeling exercises resulting in graphical images, the Huntsman team was able to visualize the outcome of each scenario and discuss those outcomes in terms of operational improvements and risk mitigation. This was a powerful set of analyses.

What benefits were derived from this approach? First, Huntsman finally had hard data points established in terms of impacts to its supply chain and S&OP process through the identification of the 13 sources of uncertainty. Second, the company identified several operational and tactical approaches they could take with minimal risk to the supply chain. This included moving inventory upstream for more flexibility, postponement, and then developing a form of risk pooling for raw materials with suppliers.

By leveraging these powerful new tools and techniques, the company was also able to reduce its planning cycle times by almost half and eliminate the "panic-reactive" efforts they relied on. The company also began to institutionalize a contingency planning approach, using the tools and techniques discussed above, into the monthly S&OP process. This allowed more scenario planning and what-if discussions to take place inside the S&OP process. The end result was better mitigation of supply chain risks. Over time, inventory was reduced by almost half and service levels improved dramatically through scenario planning and subsequent risk response plans to mitigate risks, if and when they appeared.

Combining probabilistic tools and techniques with formal risk assessment methodologies can, as demonstrated by Huntsman, effectively identify, assess, mitigate, and manage risk while improving the bottom line.

CONCLUDING THOUGHTS

Most supply chain professionals reside in a comfort zone that is populated with deterministic models and approaches. As mentioned throughout this chapter, these models have served a clear purpose over the last 50 years or so, and they will continue to enjoy widespread use and refinement. However, these tools were developed in an era where SCRM was not even an afterthought. Consequently they fail to consider the supply chain uncertainty that has become a way of life. If we can sum up the basic premise of this chapter in one phrase, it is that deterministic thinking must give way to probabilistic thinking. This will lead to the development of new approaches that emphasize uncertainty. It will also lead to the extension of existing tools and techniques where uncertainty is explicitly considered in the analysis.

Summary of Key Points

- Stochastic/probabilistic methods and models have been around for more than 50 years. Yet, supply chain management professionals are now just understanding that these tools and techniques can be leveraged to mitigate and manage risk because they overtly take into account uncertainty.
- Many of the stochastic/probabilistic tools and techniques support what-if scenario planning approaches to evaluate uncertainty, complexity, and risk within global supply chains.
- Emerging stochastic/probabilistic modeling tools are being combined with lean/Six Sigma techniques such as DMAIC as a data-driven, fact-based framework and also design of experiments to

ensure that model outcomes are statistically significant and provide sensitivity curves that explain the cause-and-effect relationships within model scenarios.

- To manage risks in complex supply chains, leading companies will combine risk response plans with digital modeling to provide powerful risk management frameworks.
- Many best-in-class companies that practice elements of SCRM are also leveraging additional lean/Six Sigma methods such as FMEA to identify, codify, classify, and force-rank risks throughout their supply chains and processes.
- Several leading companies in SCRM are also injecting stochastic/probabilistic tools and techniques along with lean/Six Sigma methods into their normal S&OP processes to not only assist in supply/demand balancing, but also to develop a risk assessment process to make informed decisions about their supply chains that take into account uncertainty, complexity, and risk.
- The methodology of leveraging digital modeling, probabilistic tools, discrete-event simulation, and risk assessment is a powerful environment to evaluate operational alternatives without experimenting on customers.

ENDNOTES

1. Accessed from http://www.apics.org/industry-content-research/publications/apics-dictionary.
2. Arntzen, Bruce, PhD Director Global Scale Risk Initative, Massachusetts Institute of Technology. *"MIT Scale Survey Presentation."* APICS International Conference 2009.
3. Accessed from http://www.apics.org/industry-content-research/publications/apics-dictionary.
4. Murray, Peter. CIRM, "Next Generation of S&OP: Scenario Planning with Predictive Analytics and Digital Modeling." *Journal of Business Forecasting,* 29, 3 (Fall 2010): 20–31.
5. Baxendell, Richard. *"Coupling Lean/Six Sigma DMAIC Methodology with Digital Modeling/Discrete-Event Simulation and DOE to Drive Profitable Manufacturing Response."* IQPC International Conference, April 2008.
6. Van Landeghem, Hendrik, and Hendrik Vanmaele. University of Ghent, Belgium, "Robust Planning: A New Paradigm for Demand Chain Planning." *Journal of Operations Management,* 20 (2002): 769–783.
7. Supply Chain Council SCOR Model, Accessed from www.Supply-Chain.org.

11

Using Big Data and Analytics to Manage Risk

By now you've probably heard about or have had some experience with something called "big data." While we may have heard of the concept, taking advantage of the treasure trove of data that resides at most companies remains an evolving challenge. With an estimate of more than 15 million gigabytes of new information collected every day (15 petabytes), which is eight times the information in all U.S. libraries, it's no wonder most companies are wondering how to use big data to their advantage.[1]

But is using big data going to be that straightforward? A report titled *Big Data Insights and Innovations Report* revealed some findings that relate directly to big data and its uses.[2] First, many organizations are challenged by data overload and an abundance of trivial information. And important data are not reaching practitioners in efficient time frames. Current technology is also not yet at the level of providing measurable, reportable, and quantifiable data in areas including production scheduling, inventory, and customer demand across the entire supply chain. Furthermore, despite the sophistication of current systems, data are not always easily accessible to internal users. Finally, noticeable gaps are present in many end-to-end supply chain flow models. Other than these "minor" issues, everything is working just fine in the world of big data and risk management.

In this chapter we'll advance some definitions and an overview of big data and predictive analytics; talk about the process for successfully leveraging big data; present barriers and challenges with big data; and present tools, techniques, and methodologies that support big data and analytics. The chapter concludes with examples of companies using big data and how these companies are leveraging their data to help manage supply chain risk.

WHAT IS BIG DATA AND PREDICTIVE ANALYTICS, REALLY?

To some observers big data got its start around 2003 with the advent of the Data Center Program at Massachusetts Institute of Technology (MIT).[3] Before this, most of the early research in the late 1990s used the term *data analytics* as a key descriptor. It becomes critical to define the terms *big data* and *predictive analytics*.

According to the Leadership Council of Information Advantage, *big data* is not a precise term. This group sees it as data sets that are growing exponentially and that are too large, too raw, or too unstructured for analysis using relational database techniques. So, where is this unbelievable amount of unstructured data coming from? According to one source, the amount of data available is doubling every two years and is emanating from not only traditional sources but also industrial equipment, automobiles, electrical meters, and shipping crates, just to name a few. The information gathered includes parameters such as location, movement, vibration, temperature, humidity, and chemical changes in the air.[4]

Predictive analytics (PA) encompasses a variety of techniques from statistics, data mining, and game theory that analyze current and historical facts to make predictions about the future. In business, predictive models exploit patterns found in historical and transactional data to identify risks and opportunities. Models capture relationships among factors to allow assessment of risk or potential associated with a particular set of conditions, guiding decision making for specific transactions.[5]

Predictive analytics has been traditionally used in actuarial science, financial services, insurance, telecommunications, retail, travel, health care, and pharmaceuticals. Yet it is barely mentioned in the manufacturing and supply chain arenas. One of the best known and early applications of PA is credit scoring, which is used throughout financial services. Scoring models process customers' credit history, loan applications, customer data, and so forth, in an effort to rank-order individuals by their likelihood of making future credit payments on time. A well-known example is the FICO score.

IBM, a leading provider of big data systems, maintains that more than 90% of the data that exists in the world today was created within the last two years. We are in an age where more than 2.5 quintillion bytes of data are created every day! We are increasingly becoming familiar with terms such as follows[6]:

- gigabytes (a unit of information equal to one billion (10^9) or, strictly, 2^{30} bytes)
- petabytes (2^{50} bytes; 1,024 terabytes, or a million gigabytes)
- exabytes (a unit of information equal to one quintillion (10^{18}) bytes, or one billion gigabytes)
- zettabytes (a unit of information equal to one sextillion (10^{21}) or, strictly, 2^{70} bytes)
- yottabytes (a unit of information equal to one septillion (10^{24}) or, strictly, 2^{80} bytes)

Don't be concerned if these definitions are confusing. They confuse us also.

IBM has been at the forefront of articulating the concept of big data.[7] In one of its analyses, the company concludes that big data, which admittedly means many things to many people, is no longer confined to the realm of technology. It has become a business priority given its ability to affect commerce in a globally integrated economy. Organizations are using big data to target customer-centric outcomes, tap into internal data, and build a better information ecosystem. IBM has created a topology that looks at big data in terms of four dimensions that conveniently start with the letter V.

The first dimension of big data is *volume*, which represents the sheer amount of data. Perhaps the characteristic most associated with big data, volume refers to the mass quantities of data that organizations are trying to harness to improve decision making. As mentioned, data volumes continue to increase at an unprecedented rate. However, what constitutes truly *high* volume varies by industry and even geography and can be smaller than the petabytes and zettabytes often referenced in articles and statistics.

Next, *variety* refers to the different types of data and data sources. This dimension is about managing the complexity of multiple data types, including structured, semistructured, and unstructured data. Organizations need to integrate and analyze data from a complex array of both traditional and nontraditional information sources within and outside the enterprise. With the proliferation of sensors, smart devices, and social collaboration technologies, data are being generated in countless forms such as text, web data, tweets, sensor data, audio, video, click streams, log files, and much more. The bottom line is that data come in many forms.

The third dimension, *velocity*, refers to data in motion. The speed with which data is created, processed, and analyzed continues to accelerate. Contributing to this higher velocity is the real-time nature of data creation, especially within global supply chains, as well as the need to incorporate streaming data into business processes and decision making. Velocity impacts latency—the lag time between when data are created or captured and when they are accessible and able to be acted upon. Data are continually being generated at a pace that is impossible for traditional systems to capture, store, and analyze, resulting in the development of new technologies with new capabilities.

Finally, *veracity* refers to the level of reliability associated with certain types of data. Striving for high-quality data is an important big data requirement and challenge, but even the best data cleansing methods cannot remove the inherent unpredictability of some data, like the weather, the global economy, or a customer's future buying decisions. The need to acknowledge and plan for uncertainty is a dimension of big data that has been introduced to executives to better understand the uncertain world of risk around big data. Veracity requires the ability to manage the reliability and predictability of imprecise data types.

A good portion of the data within global supply chains is inherently uncertain. The need to acknowledge and embrace this level of uncertainty is the hallmark of big data and supply chain risk management. An example is in energy production where the weather is uncertain but a utility company must still forecast production. In many countries, regulators require a percentage of production to emanate from renewable sources, yet neither wind nor clouds can be forecast with precision. So, what to do? To manage this uncertainty, analysts, either in energy or supply chain management, need to create context around the data.

One way to manage data uncertainty is through something called *data fusion*, where combining multiple, less-reliable sources creates a more accurate and useful set of data points, such as social media comments appended to geospatial location maps. Another way to manage uncertainty is through advanced mathematics that embrace uncertainty, such as probabilistic modeling, discrete-event simulation and multivariate, nonlinear analyses coupled with failure mode effects analysis (FMEA).

Most observers predict a major impact of big data and predictive analytics on the global economy. In a recent *Fortune* article, an expert from

Gartner suggested that over a relatively short time period, more than four million positions worldwide will emerge for analytic talent, of which only about one third will be filled.[8] Dice.com has identified the Top 10 technical skills big data will need over the next several years. By a large margin the first is Hadoop plus Java, which is not surprising since Java powers Yahoo, Amazon, eBay, Google, and LinkedIn. After that it is Developer, NoSQL, Map Reduce, BigData, Pig, Linux, Python, Hive, and Scala.

The shortage of professional skills in Hadoop and NoSQL has given rise to higher pay for qualified hires, topping $100K on average. The real winner here could be the U.S. economy. Anticipating a multiplier effect, one observer predicts that for every big data–related role in the United States, employment for three people outside IT will be created.[9] While the rise of big data presents opportunities, a shortage of qualified IT professionals also exposes an organization to risk.

As we conclude this overview of big data and predictive analytics, it would be appropriate to close this section with some key findings from IBM's research into big data. First, across multiple industries, the business case for big data is strongly focused on addressing customer-centric objectives. Second, a scalable and extensible information management foundation is a prerequisite for big data advancement. Third, organizations are beginning their pilot and implementation programs by using existing and newly accessible internal sources of data. Next, advanced analytics capabilities are required, yet often lacking, for organizations to get the most value from big data. And finally, as organizations' awareness and involvement in big data grows, four stages of adoption emerge, which the next section presents.

THE PROCESS OF SUCCESSFULLY LEVERAGING BIG DATA FOR MAXIMUM BENEFIT

Many of the cases we describe later in the chapter maintain the hallmarks of supply chain analytic implementations. These hallmarks include a clear business problem with supporting metrics; a focus on fact-based decision making and on improving business key performance indicators (KPIs); and the establishment of an end-to-end, enterprise-wide process that is championed by C-level management. Other characteristics include forward-looking scenarios and causal analysis to understand variability and performance drivers without getting lost in the data. Scenarios are performed iteratively to demonstrate value and self-fund subsequent improvement opportunities.

Many organizations start with spreadsheets as a proof-of-concept (POC) and then migrate to some form of business intelligence tool to perform more rigorous analysis. Why? According to the Hackett Group, world-class procurement organizations on average spend less than 30% of their time compiling data, compared with 60% for the bottom-quartile companies. In other words, while typical companies still compile data, world-class organizations spend more of their time analyzing the data and making informed decisions. The CIO of a leading company argues that 75% of the effort and cost when managing data is process reengineering and data cleansing and creation, and the other 25% is the IT portion. He further states that when people say their systems didn't deliver, the chances are they missed the 75% part they should have been working on.

An adoption process or continuum has emerged through observation of big data and predictive analytic projects, or what IBM calls the Four E's: Educate, Explore, Engage, and Execute." Figure 11.1 depicts this emerging adoption continuum. We'll briefly touch on the key elements of each stage.

Education is the first stage in the continuum. Its primary focus is on awareness and knowledge development. In this stage, most organizations are studying the potential benefits of big data technologies and analytics and trying to better understand how big data can help address important business opportunities. Also within the first stage, the potential for big data has often not yet been fully recognized and embraced by business executives. *Exploration*, the second stage, defines the business case and roadmap. Almost 50% of respondents in an IBM study report formal, ongoing discussions within their organizations about how to use big data

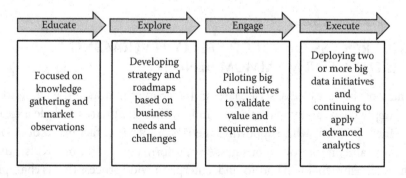

Educate	Explore	Engage	Execute
Focused on knowledge gathering and market observations	Developing strategy and roadmaps based on business needs and challenges	Piloting big data initiatives to validate value and requirements	Deploying two or more big data initiatives and continuing to apply advanced analytics

FIGURE 11.1
The Four E's of big data adoption.

to solve important business challenges. Key objectives in this stage include developing quantifiable business cases and creating a big data blueprint or roadmap. Most organizations in this stage are considering existing data, technology, and skills and are contemplating where to start and how to develop a plan aligned with their organization's business strategy.

Engagement, the third stage, is about embracing the value of the data. In this stage, organizations begin to prove the business value of big data as well as performing assessments of their technologies and skills. Companies in this stage usually have one or more proof-of-concept projects under way. These companies are working within a defined scope to understand and test the technologies and skills required to capitalize on the new sources of big data.

Execute is the final stage of the continuum. In this stage, big data and analytics capabilities are more widely operational and implemented within the organization. Many organizations here manage at least two or more big data solutions at scale, which seems to be a threshold for advancing in this stage. The companies in the execute stage are leveraging big data to transform their businesses and thus are deriving the greatest value from their information assets.

Most organizations are in the early stages of big data development. IBM's research concludes that 24% of companies are focused on understanding the concept and have not begun the journey, while 47% are *planning* big data projects and developing roadmaps. Another 28% of companies are developing proofs of concept or have already implemented full-scale solutions.

BARRIERS AND CHALLENGES MOVING FORWARD

The challenges to utilizing big data differ as organizations move through each of the four stages as featured in Figure 11.2. A consistent challenge, regardless of stage, is the ability to articulate a compelling business case. At each stage big data efforts rightfully come under fiscal scrutiny. The current global economic climate and supply chain risk landscape has left businesses with little appetite for new technology investments without measurable benefits. After companies begin their proof of concept, the next biggest challenge is finding the right skill sets to operationalize big data, including technical, analytical, and governance skills.

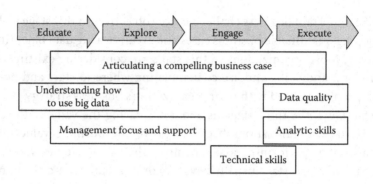

FIGURE 11.2
Big data primary obstacles. (Source: Adapted from IBM 2013 Big Data Executive Report.)

As shown in Figure 11.2, different obstacles surface as companies move through the continuum[10]:

- **Acquisition of data.** Data are available from so many sources and end users must constantly decide which will be useful.
- **Choosing the right architecture.** This involves balancing cost and performance to obtain a platform based around programming techniques far different from those of the normal desktop environment.
- **Shaping the data to the architecture.** This involves spending time capturing, compiling, and uploading the data to be aligned with the architecture. With all the new technology, transforming the data can be a time-consuming process.
- **Coding.** This includes selecting a programming language, designing the system, deciding on an interface, and being prepared for a rapidly changing environment.
- **Debugging and iteration.** This is the process of looking for errors in code, architecture, and making modifications quickly.

TOOLS, TECHNIQUES, AND METHODOLOGIES SUPPORTING BIG DATA

Let's profile the techniques that are being utilized by organizations running big data projects. Figure 11.3 gives us a sense of the tools and techniques that are being leveraged. More than 75% of companies report using core

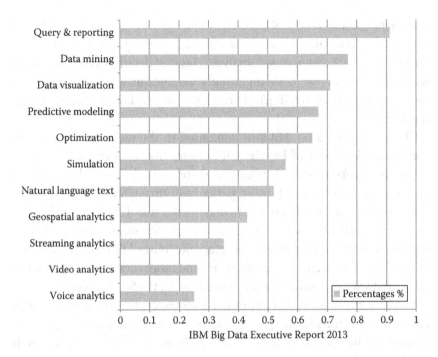

FIGURE 11.3
Big data analytics capabilities and tools.

analytics capabilities, such as querying and reporting and data mining to analyze big data, while more than 67% report using predictive modeling.

These foundational methods are a pragmatic way to start interpreting and analyzing big data. The need for more advanced visualization techniques increases with the introduction of big data because datasets are often too large for business or data analysts to view. The next highest usage of techniques and tool sets are optimization models and advanced analytics to better understand how to transform key business processes. Many organizations are embracing simulation and pattern recognition to analyze the many multivariate, nonlinear relationships within big data.

As you can see from Figure 11.3, more and more organizations are focusing on unstructured data to analyze text in its natural state, such as transcripts from call center conversations. These tools and techniques include the ability to interpret and understand the nuances of language, such as sentiment, slang, and intention. And with the tools emerging to analyze these new and unstructured forms of data, it's no surprise that the skills to manage these techniques are in short supply. It seems apparent

212 • *Supply Chain Risk Management: An Emerging Discipline*

that there are a host of tools and techniques emerging to support the big data effort. Techniques such as standard reporting, ad hoc reporting, query drill-down, cloud-based analysis, classical deterministic forecasting techniques, predictive modeling, simulation, optimization, pattern recognition, and artificial intelligence are all coming on board at an accelerating rate.

Changing gears a bit, but still remaining in the tools, techniques, and methodologies arena of big data, it appears that organizations that are embracing Software-as-a-Service (SaaS), or cloud-based technology, are utilizing those tools much more pervasively throughout their organizations, partly because they are able to make better use of scarce IT skills.[11] We mentioned earlier the lack of technical skills, in house, as an obstacle. Also, it appears that organizations that leverage outsourced IT tools and consulting skills are experiencing a much richer and more complete solution when compared with organizations that are not using the SaaS approach.

Here is a quick-hit definition of the SaaS concept: With SaaS or cloud-based business intelligence (BI), the software itself is not licensed, owned, or installed by the organization. Instead, the software resides in a remote third-party data center and the functionality provided by the software is accessed over the Internet and rented. This service is typically paid for as a monthly subscription.

One research group has concluded that the use of SaaS to drive big data analytics offers advantages across many dimensions.[12] More than 60% of organizations using a SaaS solution were satisfied or very satisfied with ease-of-use of this approach as opposed to only 41% of companies not using SaaS. Just over 80% of SaaS BI users have access to drill-down to detail capability as opposed to 58% for non-SaaS users. And just over 60% of SaaS BI users are able to tailor their solution quickly as opposed to only 41% of non-SaaS users. Companies that utilize SaaS BI tools say that they can find information they need in time to support their decisions within one hour of raw data being captured, or what is called *time-to-decision* and *time-to-value*, 84% of the time as opposed to only 70% of the time with organizations not using SaaS BI. Finally, organizations that use the SaaS approach are 40% more likely than others to exchange data openly and easily across business units. Other findings not mentioned here also reveal the value of Saas.

The idea of augmenting what you already have in house with third-party companies is gaining traction, especially with analytics. Many companies have packaged and outsourced supply market intelligence while others have developed hosted analytics tools and bundled professional services and analysis across the entire supply chain spectrum.

We'd like to close this section with some comments on analytics in manufacturing from the vice-president of Invensys Solutions. He talks about big data for manufacturing by comparing big data for consumers and the use of Google. He says we start Google maps on our phone and it immediately knows where we are. We click a box and it shows the traffic from here to the airport. If we're hungry, it can pull down restaurants and menus. Now, take this into the supply chain arena. All that information is available, but instead of restaurants, we're looking for the best batch, optimal production run, and more. A good analytics manufacturing system has to show what is out there, display the information that is available, interpret what it means, how to react to it, and then help predict what is coming next.

HOW EARLY ADOPTER COMPANIES LEVERAGE BIG DATA

With a broad definition of big data now established, we want to provide examples of industries that are leveraging big data for competitive advantage. In this section we'll touch on several industries and dig a bit deeper into a few name-brand companies that are leveraging this approach for a competitive advantage.

During our initial evaluation of the big data landscape, we speculated that there might be an industry that is far and away the leader in leveraging big data for competitive advantage. With that hypothesis, we attempted to gather data and profile the use of big data by industry. The hypothesis that one industry might dominate the landscape is far from reality. The top four industries within our sample that use big data are consumer packaged goods (CPG)/grocery (16% of firms), electronics (10%), automotive assembly (10%), and energy (10%). All other industries were 7% or less. It's evident that these industries are leading the way toward leveraging predictive analytics to solve operational problems, followed by additional industries beginning their use of big data. Overall, we have a long way to go before the use of big data becomes routinized.

Consumer Packaged Goods

A large multibillion dollar CPG company with razor-thin margins was facing highly volatile commodity prices on the supply side of its business and unforgiving, price-sensitive retailers on the demand side. The company's approach was to develop an integrated Sales & Operations Planning process and link it to the supply market by integrating supply market intelligence and purchase price forecasts. The company used this forward-looking supply intelligence to create robust scenarios, perform additional analyses, and then optimize all options associated with each scenario. It attempted to mitigate risk by finding substitute materials, modifying specifications, reconfiguring its product mix, changing its supply chain network and delivery methods where possible, hedging on the financial side, as well as modifying strategies throughout the planning horizons. This effort resulted in minimizing millions of dollars of product/customer profit erosion and more robust, predictable strategies.

Dell Computers

Most of us know that Dell is a company in transition. After dominating the enterprise PC market for decades, the Texas-based configure-to-order manufacturer is making a definitive move away from the product side of the business and toward services and solutions. Unfortunately, over the past decade, Dell's strategy, options, and variants in models, software configurations, memory, screens, and other customizable features has resulted in over seven septillion possible configurations of Dell's products! A septillion is equivalent to 1,000,000,000,000,000,000,000,000. Obviously, product portfolio complexity had become a major risk for the company.

To trim its product portfolio Dell began to utilize its abundance of big data. A Dell team created a new system called *optimized configuration*. Dell's analytics team clustered high-selling configurations from historical data to create technology roadmaps. The team also created automated algorithms to identify what configurations Dell should build to order and what Dell should produce for inventory. The analysis leveraged historical data and ran cluster analysis to identify the most common configurations sought by customers.

Clustering around commonality of product ordering allowed Dell to trim the seven septillion options to several million and provided the company's marketing and supply chain teams with agreement on preconfigured products built for inventory and ready to ship. This new supply chain strategy, driven by data analysis, also supports the company's make-to-order strategy.

Still another use of big data at Dell has been inside the company's online ordering system. Dell's business intelligence team ran analytics on click stream data, tracing every move and path taken by customers. The outcome showed that customers navigate through more than 40 clicks to place an order. The team used that information to optimize the site and reduce the number of clicks to five.

Western Digital

Western Digital, a global manufacturer of disc drives, is obsessed with quality. To serve that obsession the company has transformed its manufacturing process to allow scanning, recording, testing, and tracking of every disc drive produced while still on the production line. By running real-time shop-floor analytics, the company can locate and remove non-conforming discs before they reach the customer. Even if a disc passes an initial analytical review, if further analysis reveals a problem, the disc can be located and pulled from inventory bins. This capability, supported by big data, has resulted in the lowest warranty return rate in the entire industry. It has also helped make Western Digital the supplier of choice for many computer manufacturers.

Harley Davidson

Harley Davidson, the king of the hogs, introduced software that tracks even the minutest details on the assembly floor, such as the speed of fans in the painting booths. When the software detects that the fan speed, temperature, or humidity has deviated from the optimal settings, it automatically adjusts the operations. This allows a consistency on the shop floor by staying within preestablished parameters. The software has also been used to identify bottlenecks on the assembly floor. One of Harley's goals is to complete a motorcycle every 86 seconds. A recent study using shop floor data revealed that the rear fender assembly time was taking longer than planned. The company changed the factory configuration so the fenders would flow directly to the assembly line rather than being placed on carts and moved across the floor. This is but one example of how

Harley Davidson is using big data to streamline its operations and avoid operational risk.

Raytheon

Raytheon, a household name in the aerospace/defense manufacturing arena, is betting on big data to reduce the risk of quality and operational problems. In its Huntsville, Alabama, missile plant, if a screw is supposed to be turned 13 times after it is inserted but instead is turned only 12 times, an error message flashes and production of the missile or component stops. "Manufacture of a missile is either right or it's not; there's no in between," says a Raytheon executive. Many manufacturers are installing sophisticated automated systems to gather and analyze shop floor data, known as manufacturing execution systems (MES). Manufacturers are looking harder at data partly because of increasing pressure from customers to eliminate defects and from shareholders to squeeze out additional cost and mitigate risk to the brand. These new capabilities mean Raytheon is catching more flaws as they occur. Raytheon also keeps data for each missile, including the names of all the machine operators who worked on any part of it, as well as the humidity, temperature, and more at each workstation.

The system is designed to prevent any operator from performing a task for which he or she is not certified. According to Raytheon, leveraging big data systems is a form of risk mitigation and management. Millions of dollars have been spent in the past to rework items that did not meet specifications. If Raytheon's experience is any indicator, cost containment, real-time event monitoring, and process optimization are but a few of the key drivers supported by big data. Tracking physical items and people throughout the supply chain, capturing and acting on streaming data, and enabling faster reaction to specific problems before they escalate in major situations is becoming the norm rather than the exception.

European Electrical Utility

A major European electric utility company sought to improve the management of budgeted versus actual spend for nonfeedstock and indirect spending. It wanted a single system that separated consumption variation, within a contract and across contracts, external market pricing variation, and procurement-led pricing impacts. The company used a third-party tool for spend analysis and procurement performance analysis. These data were cross-referenced against an external database with thousands of price indexes. This approach of combining internal and external data and benchmark indexes allowed for fact-based discussions and decision making for continuous improvement in its cash flow management.

This approach confirmed that analytics works best when integrated with external information. The utility company concluded that integrating internal and external data through the use of big data analytics is an enabler for managing supply risk, supplier risk, regulatory risk, competitive risk, and intellectual property risk. Managing these risks should include analytical approaches such as scenario planning, Monte Carlo/ probabilistic modeling to quantify the probability of occurrence and impact, segmenting and visualizing risk using heat maps, and predictive analytics to manage risk.

Schneider

Schneider National, a $3 billion transportation and logistics company, has developed a computer model that mimics human decision making, helping the company to assign trucks and drivers in the most cost-effective way possible. At any given time, Schneider has 10,000 trucks on the road with over 30,000 trailers waiting to be picked up or delivered. Drivers work alone or in pairs, and Schneider must get them back home by a certain date and time. Drivers also need to conform to the government's hours-of-service regulations regarding rest periods and breaks.

With the help from several Princeton University researchers, Schneider developed a simulator utilizing dynamic programming, which takes into account the presence of uncertainty. The simulator, which took two years to develop, runs forward in time for three weeks to approximate the value of having a truck and driver at a certain location at a certain time. The output from the report is a called a first-pass cost estimate. The tool then runs backward in time, something called postanalysis, to reconcile the past results with those that were determined in the future estimate. The simulator then runs forward again for three weeks and then backward as it seeks to improve the total cost estimates. This forward–backward process encompasses hundreds of thousands of iterations.

Schneider estimates its big data tool has saved the company tens of millions of dollars as well as increased revenue by justifying price hikes to customers with specific service-level constraints. The simulator also provides a quantifiable methodology to exercise what-if analyses, such as determining the marginal value of hiring new drivers in a certain region to handle growing volume. Moving forward, Schneider expects to use the simulation tool to help identify new businesses to pursue. As an operations research analyst at Schneider says, "The tool is so powerful that when someone presses us on the impact of different customer policy changes, we have the facts and we have the data. The value of this big data tool is to be able to take these complex business issues and opportunities and give them a good, solid analysis."[13] Schneider understands clearly the link between data management and strategic risk management.

CONCLUDING THOUGHTS

Few should disagree with the notion that big data and predictive analytics are here to stay. Big data, predictive analytics, and many of the tools and techniques discussed in this chapter and in Chapter 10 are providing approaches for codifying, classifying, analyzing, and acting on vast amounts of data, most of which is the size many of us have never dealt with in our professional careers. We've witnessed an increasing number of companies that have leveraged their data and predictive analytics to solve complex supply chain, manufacturing, and customer-centric issues to enhance revenue, reduce cost, improve asset utilization, and reduce supply chain risk.

A study by the Aberdeen Group reinforces the finding that top performers are making advanced analytics activities a priority to take control of manufacturing complexity and supply chain risk. This is a good thing since uncertainty, complexity, and risk continue to grow globally. Research reveals that the companies that are best at leveraging big data average a 19% year-over-year increase in operating profit as opposed to only a 9% increase for all other companies. And 80% of companies that are the best at leveraging big data have witnessed improvement in the cycle times of their key business processes over a one-year period, as opposed to 47% for average companies and 39% for laggard companies.[14] Increasingly, the ability to compete successfully as well as manage supply chain risk will rest upon a company's ability to leverage big data and predictive analytics.

Summary of Key Points

- The Leadership Council of Information Advantage sees big data as data sets that are growing exponentially and that are too large, too raw, or too unstructured for analysis using relational database techniques.
- Predictive analytics (PA) encompasses a variety of techniques from statistics, data mining, and game theory that analyze current and historical facts to make predictions about the future.
- IBM has created a topology that looks at big data in terms of four dimensions: volume, variety, velocity, and veracity.
- One way to manage data uncertainty is through data fusion, where combining multiple, less-reliable sources creates a more accurate and useful set of data points. A second way is through advanced mathematics that embrace uncertainty.
- World-class procurement organizations on average spend less than 30% of their time compiling data, compared with 60% for the bottom-quartile companies. World-class organizations spend more of their time analyzing the data and making informed decisions.
- An adoption process or continuum has emerged through observation of big data and predictive analytic projects, or what IBM calls the 4-Es: Educate, Explore, Engage, and Execute.
- Foundational methods are a pragmatic way to start interpreting and analyzing big data, but the need for more advanced visualization techniques increases with the introduction of big data because datasets are often too large for business or data analysts to view.
- Consumer packaged goods (CPG)/grocery, electronics, automotive assembly, and energy are the four industries leading the way toward leveraging predictive analytics to solve operational problems.

ENDNOTES

1. McKendrick, Joe. "Big Data, Big Issues: The Year Ahead in Information Management." 2010.
2. Eshkenazi, Abe. APICS Big Data Insights and Innovation Report, 2012.
3. Schuster, Edmund. "Big Data Is a Big Reality." Accessed from http://ingehygd.blogspot.com/2012/02/big-data-is-big-reality.html.
4. Lohr, Steve, "The Age of Big Data." *New York Times*, accessed from http://www.nytimes.com/2012/02/12/Sunday-review/bid-data-impact-in-the-world.html.

5. Accessed from http://www.apics.org/industry-content-research/publications/apics-dictionary.
6. All definitions were retrieved from Google.
7. IBM Global Business Services Business Analytics and Optimization Executive Report, "Analytics: The Real-World Use of Big Data." in collaboration with Said Business School at the University of Oxford, 2012–2013.
8. Fisher, Ane. "Big Data Could Generate Millions of Jobs." *Fortune*, May 21, 2013.
9. Fisher, Ane. "Big Data Could Generate Millions of Jobs." *Fortune*, May 21, 2013.
10. Crandall, Richard."The Big Data Revolution." *APICS Magazine*, March/April, 2013.
11. White, David. Aberdeen Group's Analyst Insight Report, "Software-as-a-Service Helps Deliver Satisfied Analytics Users." May 2013.
12. White, David. Aberdeen Group's Analyst Insight Report, "Software-as-a-Service Helps Deliver Satisfied Analytics Users." May 2013.
13. Coster, Helen."Calculus for Truckers." *Forbes*, 2013.
14. Lock, Michael. Aberdeen Group's Embedded Analytics Report, March 2013.

12

Emerging Risk Management Tools, Techniques, and Approaches

This chapter presents a variety of emerging tools, techniques, and approaches to address and manage supply chain risk. They appear here because they are not fully developed or implemented at most firms. The approaches we present include becoming a preferred customer, creating supply chain heat maps, mapping the supply chain, declustering clusters, creating a flexible supply chain, creating a risk war room, and managing operational working capital. Other emerging approaches, including total cost measurement, estimating available supplier capacity, and calculating risk scores, appear elsewhere in the book.

BECOME A PREFERRED CUSTOMER

Supply chain leaders understand that the link between positive relationships with suppliers and improved corporate performance is a strengthening rather than weakening one. Companies that fail to develop positive relationships may find their suppliers allocating limited capacity to other firms, sharing their most innovative ideas with other customers, or exiting an industry segment altogether, all of which will lead to increased risk. As our research clearly shows, becoming the preferred customer to suppliers offers advantages that are not as readily available to other customers, advantages that could lead to future competitive advantage and reduced risk.[1] As one executive summed up clearly during a research interview, "Becoming a preferred customer is going to be one of the best ways we have to manage supply chain risk in the future."

A research project involving hundreds of suppliers revealed a clear link between a customer's (i.e., the buyer) behavior, the satisfaction level a supplier has with a customer, and a supplier's willingness to provide preferential treatment that less-satisfied suppliers are not willing to provide. The following summarizes three important findings from that research.

Supplier satisfaction relates directly to a customer's performance and behavior rather than demographic or other attributes.
Supplier satisfaction correlates significantly with factors that relate to a customer's behavior toward the supplier (i.e., pay on time, share relevant information, treat suppliers ethically, etc.) rather than demographic or other factors such as supplier size or the size of a contract. Interestingly, a slight negative correlation exists between the total years a supplier has worked with the customer and lower satisfaction with that customer, which presents clear risk implications.

No statistical relationship exists between the size of a supplier in terms of sales and supplier satisfaction with the buying customer. Furthermore, no relationship exists between the size of the contract relative to the supplier's total sales and supplier satisfaction with the buying customer. Supplier satisfaction relates directly to customer performance and behavior rather than the volume of sales that the customer represents. This is welcome news because it suggests customers can change their behavior to improve supplier satisfaction, thereby leading to preferred customer benefits.

The relationship between supplier satisfaction and viewing a customer as preferred is extremely strong.
The correlation between supplier satisfaction and viewing a customer as preferred is unusually strong. This strong indicator reveals a clear link between satisfaction and preferred customer status. An important conclusion is that becoming a preferred customer will likely not occur if a supplier is dissatisfied with a buying customer.

Satisfied suppliers are more willing to provide valuable kinds of preferential treatment to their preferred customers compared with less-satisfied suppliers.
A clear statistical relationship exists between supplier satisfaction with a customer and the willingness of that supplier to provide certain (and valuable) kinds of preferential treatment. What is surprising is that in

Supplier-Provided Direct Investment	Supplier-Provided Innovation	Supplier-Provided Favorable Treatment
• Capacity dedicated to the customer • Personnel to work directly at the customer's facilities • Engineers to support the customer's product design needs • Investment in new equipment that benefits only the customer • Exclusive use of new technology developed by the supplier • Hold inventory to support the customer's needs • Provide direct financial support if needed • Create information technology systems unique to business with the customer	• Product innovation • Production process innovation • Process and supply chain innovation other than production processes	• Shorter quoted lead times • Preferential scheduling of orders • Early insight into the supplier's future product technology plans • More favorable payment terms • Performance improvement ideas • More frequent deliveries • Access to the supplier's executive level personnel • Access to supply market information the supplier may possess • Better pricing • First allocation of output if supplier capacity is constrained • Early warning to potential supply problems

FIGURE 12.1
Supplier-provided preferential treatment.

many cases the preferential treatment that satisfied suppliers are most willing to provide to their preferred customers are the kinds of treatment that suppliers overall are *least* willing to provide. Preferred customer status brings with it preferential treatment that is not available to a typical customer, benefits that can contribute to a hard-to-duplicate competitive advantage and reduced supply chain risk. Figure 12.1 identifies the supplier-provided outcomes that suppliers may potentially provide to their most-preferred customers.

Gaining Preferred Customer Status

We know that gaining preferred customer status can bring with it some hard-to-duplicate advantages that have clear risk management benefits. The question becomes how a buying company can become a preferred customer to suppliers. The following provides some guidance about how to achieve preferred customer status.

Understand How Suppliers Perceive Your Company as a Customer. How can a buying company improve its relationships and receive preferential treatment from suppliers if it does not know in what areas it is doing well and where it is falling short? Gaining this insight requires a commitment from the highest levels of the supply organization to solicit supplier-provided feedback. Companies often use third parties to collect information from suppliers to ensure the integrity of the process.

Pursue Trust-Based Relationships with Suppliers. Few would question the importance of trust within a buyer–seller relationship. Industrial customers can demonstrate their trustworthiness through open communication with suppliers, following through on promises and commitments, sharing relevant supply chain information, and acting legally and ethically in all business dealings. The importance of ethics and protecting proprietary information within a trust-based relationship is also critical. One outcome from a customer's relationship efforts should be the pursuit of activities that promote frequent contact, particularly since communication frequency and the level of trust within a relationship are highly correlated.

Recognize the Importance of Knowledgeable Personnel and a Stable Workforce. Our research with suppliers reveals that more than 90% of suppliers agree it is "very important" that the personnel they deal with at their customer are knowledgeable. Maintaining supply knowledge will be a challenge, for example, as the baby boomer generation exits the workforce. This exodus will require a set of talent management strategies that focus on acquiring and then retaining personnel with the right set of capabilities, including knowledge about how to manage supply chain relationships and risk.

Avoid the Seven-Year Itch. Various research studies have concluded that the satisfaction that suppliers have with their industrial customer tends to decline the longer the supplier has worked with that customer. The inflection point for this downward shift often occurs around the seven-year point of the relationship. The challenge becomes one of recognizing that this downward shift is a real possibility and then developing an action plan to ensure this shift does not occur. If a downward shift has occurred, the challenge becomes one of reenergizing the buyer-seller relationship.

Request Preferential Treatment. We know that satisfied suppliers are willing to provide preferential treatment to their most favored customers. The task becomes one of understanding how to obtain that treatment. One time to address this subject is during contract negotiations, particularly when crafting a supplier's statement of work. Another time is during annual supplier review meetings, meetings that suppliers recognize as extremely valuable. Suppliers also indicate a strong willingness to engage in various forms of executive-to-executive interaction and participate in buyer–supplier councils that feature executive-level engagement between the customer and key suppliers.

Tap into Supplier Innovation. A variety of ways exist to tap into supplier innovation. These include early supplier involvement during product and technology development, technology demonstration days where suppliers are encouraged to showcase their new ideas, and supplier participation on customer improvement teams. If innovation is the lifeblood of growth, then it becomes a competitive necessity to tap into sources of innovation wherever they exist.

Never discount the important relationship between being a preferred customer to suppliers and reduced supply chain risk exposure. The statistical linkages between supplier satisfaction, preferred customer status, and preferential treatment are unambiguous.

CONSTRUCT SUPPLY CHAIN HEAT MAPS

As the concept of supply chain risk management (SCRM) began to solidify into a body of knowledge and then courseware, several major beliefs evolved into a set of questions-of-discovery regarding the maturity of the supply chain. As these tenets began to solidify, a mechanism to review, evaluate, and position a company's maturity in supply chain excellence and design took shape in terms of a *spider diagram*. This diagram profiles a company's supply chain maturity, which then relates to the risk associated with that maturity. This has evolved into a supply chain risk awareness *heat map*. The heat map involves a set of about 100 questions-of-discovery across 10 areas in an end-to-end supply chain. The answers provide a glimpse into a company's exposure to supply chain risk. The spider diagram provides a relative profile of red (high), yellow (medium), and green (low) levels of risk.

Heat map users can download a cloud-based tool and answer a set of self-assessment questions. The resulting heat map provides, potentially for the first time ever, a view into a company's risk across its entire supply chain. Some of the areas evaluated within the map include leadership, balance scorecards, information technology, sales and operations planning (S&OP) processes, supply chain techniques, supply base, and logistics. The heat map is an awareness technique that starts a dialogue about supply chain risk. Part of its value is that many companies do not maintain an integrated approach to supply chain management. These companies still have a tremendous amount of suboptimization within silos of the organization and a lack of supply chain visibility. This tool is used to establish an enterprise-wide picture of a company's supply chain and risks through questions-of-discovery answered by representatives from multiple disciplines. The appendix of this book provides more detailed insight into this powerful and emerging approach to SCRM.

MAP THE SUPPLY CHAIN

A supply chain map is a graphical representation of a firm's tier-one and subtier suppliers for any purchased item, whether it is a finished product, system or assembly, or component. Better maps will also include the downstream portion of the supply chain. Intuitively, supply chain mapping makes sense from a number of perspectives. Unfortunately, just because something makes sense does not mean it will be performed. Mapping is still an emerging risk management technique at most firms. And those that do map their supply chain usually map only a portion of it.

No standard nomenclature specifies what a supply chain map should contain or look like. This is in contrast to project management or lean management, where the tools and techniques are well established. Different kinds of maps exist, many of which are best described as the homegrown variety. Because SCRM is an evolving discipline, it will take time to converge on a set of tools and approaches, something that is characteristic of mature disciplines, or for third parties to enter the market with mapping products.

A mapping technique that is probably the most frequently used is the links and nodes approach. Links and nodes is a fairly common approach when designing or modeling a supply chain network. Some companies simply apply this technique to their mapping efforts. Nodes are the entities within a supply chain, such as suppliers, distributors, customers, other channel members, and the focal company. Links represent flows and are represented by solid or dotted lines. Flows include material, services, funds, and information.

Challenges When Mapping a Supply Chain

The growth of supply chain mapping as a risk management technique faces various challenges. For companies with a large supply base or complex product structures, the maps become intricate very quickly. A supply base with hundreds of tier-one suppliers will have thousands of tier-two and tier-three suppliers as part of the chain. The same holds true for the downstream supply chain entities as we move toward the final customer.

Another challenge is that many companies are not that concerned with what transpires past their first-tier suppliers. This is due partly to benign neglect and the fact that supply chain mapping is an intensive work process. A third challenge is that some suppliers are not willing to release their bill of material data to customers or reveal their supply sources, making it difficult to identify subtier suppliers.

A fourth challenge involves determining the right kind of maps to develop. Some maps are so high level, such as industry-level maps, that they lack the insight required to use them for effective risk management. At the other extreme, a supply chain map may contain so many connections and nodes and so much information that they become difficult to understand. The map may look like something an energetic 3-year-old created using a box of crayons. A final challenge is that supply chains constantly evolve. Like total cost models, a topic addressed in Chapter 13, supply chain maps must be kept up-to-date. This requires a disciplined work effort.

Supply Chain Mapping Guidelines

The point of this discussion is not to show how to create detailed supply chain maps. Whatever specific technique we feature could always be replaced with another technique. Rather, our objective is to present a set of guidelines, in no particular order of importance, that will help when developing supply chain maps.

- **Don't forget the demand chain.** Most supply chain maps focus on suppliers because the procurement group tends to be responsible for developing supply chain maps. Why not use the map to support supply chain integration efforts between the supply and demand sides of the supply chain? It is difficult to come up with a logical reason why companies should ignore the demand side of the supply chain when developing maps.
- **Include geographic locations.** Similar to the argument presented in the previous paragraph, it is difficult to come up with a logical reason why companies should not include the geographic locations of supply chain entities. These are, after all, called *maps*. The next emerging

approach presented in this chapter, cluster analysis, will benefit greatly from the availability of supply chain maps that include geographic locations.

- **Suppliers and sites should be different entities in supply chain maps.** Supplier corporate locations and shipping locations may be different entities. In fact, a single supplier may use numerous sites to support a buyer's operations. When this is the case, special designations should treat sites as a separate entity from the corporate entity. Also, logistical providers and their sites should be part of the map.

- **Insert hyperlinks in the map to connect to additional information and data.** A major debate with maps centers on their complexity and detail. Hyperlinks will help simplify maps while providing access to detailed data and information about supply chain entities. These links may take a user to Dun & Bradstreet reports, commodity or supplier databases, alerts, supplier performance history and scorecards, and third-party reports, to name a few possibilities.

- **Use a product's bill of material (BOM) as a guide when developing supply chain maps.** A bill of material is essentially a map of a product's structure and "recipe." BOMs provide an excellent platform from which to extend that information into a supply chain map. In a perfect world, internal users would be able to click an item on a product's bill of material and see a fully developed supply chain map.

- **Include more than material flows in the map.** Supply chain maps should also consider information and financial flows. This can be accomplished by different designations that represent linkages within the supply chain. For example, the flow of funds may be represented by a dotted rather than solid arrow.

- **Use college interns when developing maps.** Given the workload required to develop supply chain maps, it makes sense to remove the day-to-day work effort from the main flow of an organization's work.

- **Don't ignore interrelationships among firms in supply chain maps.** A supply chain entity may be part of a supply chain map for another product or business unit within your company, which the map should consider. Or, your customer's customer may also be a critical supplier. These interrelationships are what turn supply chains into supply networks.

- **Gain visibility to subtier supply chain entities.** Insisting on visibility to a tier-one supplier's suppliers should be a contracting requirement, assuming the buyer has some semblance of power over the supplier. Almost by definition, world-class companies would never relinquish their understanding of the origin of components or other items that make up their final product. Apple, for example, maintains clear visibility and control of its extended, multitier supply chain without owning any of it.[2]

- **Share the work burden.** Part of the evaluation process of tier-one suppliers can be an assessment of how well these suppliers map their supply chain. And if your tier-one suppliers insist that their tier-one suppliers (your tier-two suppliers) map their supply chain, you will gain visibility well into your supply chain downstream. In reality, this level of supply chain integration rarely happens this way. That is why major parts of the supply chain mapping process are still emerging.

DECLUSTER THE CLUSTERS

Anyone who follows the work of Harvard's Michael Porter appreciates the connection between business or industry clusters and competitive advantage. Porter defines a cluster as geographic concentrations of linked businesses that enjoy unusual competitive success in their field. Compared with market transactions among dispersed buyers and sellers, the repeated exchanges between firms in a cluster promote trust, interdependence, coordination, innovation, and communication.[3] Clusters affect competition by increasing the productivity of companies in the area, driving the direction and pace of innovation, and stimulating the formation of new businesses, which expands and strengthens the cluster.[4] In the supply chain space we also have logistics clusters, which are defined as communities that bring together a broad range of supply chain services and deep expertise.[5]

Many examples of clusters exist—movie production in Hollywood, high-technology companies in Silicon Valley, the domestic U.S. automotive industry in Detroit, business process service providers in India, and financial services in New York City. And for those of us who might need a little touch-up, hundreds of plastic surgery clinics have clustered around the "beauty belt" in an upscale Seoul, South Korea, neighborhood. Service providers around this cluster also provide special services for visitors, including hotel accommodations, airport pickups, and multilingual websites.[6]

From a risk management perspective it makes sense to think about declustering the clusters. What does *declustering* mean? Perhaps foremost, declustering requires having a solid knowledge of the location of the most important entities in your supply chain, including critical subtier suppliers. One approach is to color code the different nodes in a supply chain. Subtier supplier, first-tier suppliers, internal manufacturing sites, contract manufacturers, ports and other logistical centers, and distributors will each have a color assigned. Next, create a color-coded global map. Are your imported items arriving through a single port? Do all of your exports move through the same logistical network? Are a large number of your suppliers clustered together geographically? Are you relying on a single distributor location? Are your clusters located near areas known for hazard risk, including earthquake fault lines, tornadoes, flood plains, and hurricanes? A buying company may have its own supplier cluster simply because it relies on suppliers that are located in the same general vicinity.

The next step in the declustering process is to develop risk management plans that address any cluster-related risks. This can involve creating flexible supply chains (a concept that is discussed shortly), requiring suppliers to have multiple production locations that are not located near each other, qualifying multiple suppliers, and relying on more than a single logistical center for processing imports and exports. A well-known U.S. electronics company conducted a cluster analysis and realized it was using a single port in China to process all of its Chinese imports and a single port in the United States to receive those imports. This company now relies on more than one Chinese and one U.S. port.

Clustering Gone Wild

A prime example of the downside of clusters involves the massive flooding that occurred in Thailand, a region that featured 45% the world's hard drive production. While clusters certainly have redeeming qualities, less mentioned is their potential dark side. One only has to look at Thailand and the flooding that essentially wiped out a large portion of the electronics hard-drive industry. Chapter 5 examined this catastrophe from a hazard perspective.

Before profiling the financial impacts of the flood, let's talk about the back story. At some point multinational companies concluded, after taking into account labor, taxes, freight, logistics, and government incentives, that costs in Thailand's made it an appealing location. If someone had taken the time to develop a complete supply chain map with a 100-mile radius, he or she might have noticed that many suppliers, particularly in the computer hard drive industry, were located in a 100-year flood plain. Actually, some must have recognized this minor detail since some supplier facilities were built 6–10 feet above the flood plain. Although those plants did not get flooded directly, it became impossible to move material in and out of these facilities.

The impact from this flood was, unfortunately, impressive in its scale. More than 660,000 people were forced out of work and 9,859 factories were shut down. In the auto industry, 6,000 automobiles per day were not built. In the camera industry, Nikon eventually lost $786 million in sales; Canon had $604 million in lost sales; and Sony had recovery costs of $107 million in a single quarter. And in the computer hard drive industry, iMac launches were delayed and hard drive prices skyrocketed by 100%.

Many foreign companies, having witnessed firsthand the risk of locating too much of their production and supply chain in one location, have shifted production to countries other than Thailand. Western Digital Corporation, a U.S. company that is one of Thailand's largest employers, moved some hard-disk drive component manufacturing to Malaysia. The company also asked some of its suppliers to take similar steps. Nidec, a company based in Japan, shifted part of its production of hard-disk drive motors to China and the Philippines.[7] These are not isolated examples.

CREATE A FLEXIBLE SUPPLY CHAIN

We may not realize it yet but a concept called *flexibility* may become our new best friend when managing risk. It is certainly a concept that is receiving attention in the popular press as well as from supply chain managers. A Global Supply Chain survey conducted by PricewaterhouseCoopers (PwC) concluded that 64% of respondents plan to implement greater flexibility to better respond to supply chain challenges, making flexibility a top supply chain priority.[8]

Even though flexibility has been discussed for some time, it is still an evolving idea across the entire supply chain. Most perspectives of flexibility view this concept in terms of adjusting volumes in a manufacturing environment. Consider a definition of *flexibility* from an online business dictionary: flexibility is the ability of a system, such as a manufacturing process, to cost-effectively vary its output within a certain range and given time frame.[9] Supply chain flexibility involves much more than what this limited perspective conveys.

What is flexibility and why is it valuable? To be flexible means to be adaptable and responsive to change, including changes brought about by risk events. The opposite of being flexible means to be rigid and unable or resistant to change or modify a course of action. Consider the following example of inflexibility. In 2013 a derailment of an eastbound passenger train in Connecticut and subsequent collision with a westbound train severely affected the busiest passenger rail corridor in the United States.[10] Part of the reason the impact was so severe was that construction in the area of the accident reduced the number of active tracks from four to two over a seven-mile stretch. The result was a complete lack of routing flexibility that prohibited the diversion of trains to other tracks. Hundreds of thousands of commuters were forced to shift from the rails back to the roads, a part of the country that already experiences traffic gridlock.

Flexibility provides choices, and choices directly support greater supply chain resiliency. Flexible supply chains are able to adapt quickly to changes or risk events, including disruptions to supply and changes in demand, while maintaining customer service levels.[11] The key point here is to be able to respond quickly, sometimes within minutes of an event.

Examples of Flexibility

No single type of flexibility exists or applies to all situations. Table 12.1 describes various kinds of flexibility with ways to promote that flexibility. While this talk about flexibility sounds good, where do we see it in the real world? Following are examples of the value of supply chain flexibility.

Energy Flexibility. An increasing number of utilities operate power-generating facilities that use coal or natural gas as their energy source. As market prices shift between coal and gas, some utilities take advantage of these price changes by shifting their fuel source. One observer notes that "natural gas prices can't really shoot much more above a certain level because then utilities will pile back into coal."[12] Of course, the inverse is true.

TABLE 12.1

Supply Chain Flexibility

Type of Flexibility	The ability to...	Supply Chain Tactics
Volume flexibility	Adjust order volumes internally and with suppliers in response to changes	• Overtime and weekend production • Access to temporary labor • Contract manufacturers and secondary suppliers • Safety inventory
Order lead time flexibility	Have variable rather than fixed lead times with suppliers as required by customer demands	• Ask for shorter lead times from suppliers • Negotiate variable lead time requirements with suppliers • Select suppliers that have lead-time flexibility capabilities
Scheduling flexibility	Adjust production and delivery schedules internally and with suppliers	• Real-time data visibility and dynamic scheduling • Ask for preferential scheduling treatment from suppliers
Product configuration and variety flexibility	Modify the design of a base product, including adding new varieties or features	• Develop platform products that allow re-configurability and modification
Physical flexibility	Increase square footage or layout through physical infrastructure changes	• Use modular facilities that can be modified for new uses • Build in future expansion and re-configurability capabilities during facility design
Capacity flexibility	Modify the internal and external capacity levels of supply chain members	• Reconfigure work cells to shift according to product mix requirements • Use overtime and weekend production • Approve secondary supply sources and contract manufacturers • Reserve capacity slots with suppliers
Design flexibility	Modify product designs quickly	• Computer aided product designs • Virtual simulation and testing • Use standard components wherever possible
Internal routing flexibility	Alter how a product flows through a facility	• General rather than specialized workers and equipment • Preapproved alternate routing
Logistics flexibility	Reroute or adjust movement through logistical networks and modes of transportation	• Approved secondary carriers • Multiple port options • Multiple modal choices • Control material title of goods

continued

TABLE 12.1 (continued)

Supply Chain Flexibility

Type of Flexibility	The ability to...	Supply Chain Tactics
Source/ location flexibility	Shift production from one internal or external supplier or site to another supplier or site	• Multiple internal production sites • Multiple qualified suppliers • Suppliers with multiple production sites
Workforce flexibility	Assign and reassign workers as needed	• Simplify labor work rules and job classifications • Gain access to temporary labor
Energy flexibility	Shift seamlessly between energy sources	• Buy flex-fuel vehicles • Consider energy flexibility when specifying new equipment and facilities
Material flexibility	Substitute one component, material, or material grade for another	• Preapprove material substitutes during product design • Avoid custom designed materials
Supply chain flexibility	Shift between supply chains to accommodate different products and customer channels	• Create different supply chains for different production strategies and market segments

Coal prices cannot rise too high because utilities will shift to gas. Utilities are such large consumers of both raw materials that industry observers expect coal and natural gas prices to move inversely rather than in unison as utilities adjust their usage in response to market prices. Energy flexibility provides a natural hedge against the financial risk of rising energy costs. We expect shifts in market share between coal and natural gas to continue as utilities adjust their raw material usage based on prices.

Capacity Flexibility. Companies are increasingly pursuing longer-term contracts with suppliers that feature reserved capacity that is not committed to any particular product. As demand becomes better known, these capacity slots are replaced with actual orders. This approach provides greater flexibility, shorter lead times, assurance of supply, and of course reduced supply chain risk. The medium to longer-term forecasts that are used to reserve capacity slots usually involve an aggregate number rather than specific products.

Physical Flexibility. When a leading medical technology company that designs and manufactures orthopedic products planned its Mahwah, New Jersey, facility, it did so with flexibility as a primary objective. The facility includes work cells that can easily be reconfigured from one product line to another, a capability that allows the company to better match demand and supply across its many products. Furthermore, the building was constructed so it could be physically expanded if more square footage was required. This compares with facilities that are fixed due to proximity to other structures, roads, railroad tracks, etc.

Design Flexibility. Common wisdom would say that it is not feasible to build servers in the United States. Someone should have told that to SeaMicro (now part of AMD), a company located in Santa Clara, California. The company uses a contract manufacturer located one mile away to build a radically different kind of server. SeaMicro says that manufacturing locally has helped it compete successfully against much larger rivals. The company's engineers are constantly experimenting with new components in a bid to reduce energy consumption and quicken the performance of their servers. Engineers take their changes a mile down the road to the contract manufacturer and test them in new systems almost immediately. Design flexibility, which SeaMicro calls "lean engineering," saves weeks and even months compared with using contract manufacturers located in China or Taiwan. [13]

Product Configuration Flexibility. Over the last decade John Deere has been tearing up more than dirt. The company has also been tearing up sales as growth continues from overseas operations. While a number of factors contribute to this growth, something we cannot ignore is the company's ability to offer an unbelievable variety of product combinations. Farm equipment is not a product like aircraft, that is generally standardized country to country. In a recent year, Deere built almost 8,000 variations of its popular 8R tractor line. Product design flexibility allows Deere to serve the needs of diverse farm markets using a single product platform.

Lead Time Flexibility. An example of a company that owes a large part of its success to flexibility is the Spanish retailer Zara, a company that is synonymous with fast fashion. Since its inception, the company's founder has insisted the retailer always follow two basic rules—store inventory must be replenished twice a week and stores must receive their orders within 48 hours.[14] This allows Zara to be responsive flexibly to demand shifts and customer preferences, particularly as the retailer expands globally. We are also seeing a resurgence of small apparel manufacturers in the United States as customers place orders (usually smaller orders) with suppliers that can respond in days or weeks rather than months.[15] And, these customers do not have to worry about the risk that comes from being associated with foreign apparel factories that have abhorrent working conditions.

Supply Chain Design Flexibility. Dell Computer, a company that faces strategic risk as customers shift from PCs and laptops to tablets and other devices, realized that the responsive supply chain it established to support make-to-order online sales was not necessarily the right supply chain to support its expansion into retail sales and other market segments. This "one size fits all" supply chain did not fit Dell's needs at all. Dell has since developed four supply chains, each dedicated to a different customer segment, that give the company added flexibility to respond to a broader array of market opportunities. The build-to order supply chain supports Dell's online customer segment; the build-to-plan supply chain supports the retail segment; the build-to-stock supports the online/popular configurations segment; and the build-to-spec supply chain supports the corporate segment.[16]

Logistics Flexibility. Logistics flexibility means being able to adjust the route taken to move goods, funds, or information between points or provide choices across different modes of transportation. A $2 billion pipeline (clearly a fixed, inflexible mode of transportation) designed to take plentiful crude oil from West Texas to California is failing to generate interest among large California refiners because of the flexibility offered by rail cars. Relying on rail shipments to transport oil allows refiners to source from different locations around the United States and route the oil to their California refineries, something that is not feasible with a fixed pipeline. A growing supply of North American crude oil is coming from various locations where prices fluctuate, allowing refiners to use different routes and modes of transportation (i.e., rail cars) to make opportunistic purchases for their crude supply.[17]

Material Flexibility. In the not-too-distant past the price of nickel soared, making it a cost-prohibitive item for companies that rely on stainless steel 318, an industry standard item that contains nickel. One industry that was hit particularly hard includes companies that manufacture vehicles to carry food products. Material engineers were able to shift to "lean duplex," a type of stainless steel that offers material properties that are 30% to 200% better than traditional alloys with only a fraction of the nickel contained in other stainless steels. Lean duplex also offers higher yield strength, making it less susceptible to cracking and corrosion.[18] Material flexibility helped the producers of tank trailers avoid financial risk from commodity volatility.

CREATE A RISK WAR ROOM

Close your eyes and think about walking into a room where risk management information is collected, categorized, analyzed, prominently displayed, and widely disseminated to the right people at the right time. Imagine having a dedicated staff tasked with the responsibility for monitoring supplier health, collecting and analyzing third-party data, spotting disruptive weather patterns, tracking material movement around the globe, updating in real time a dashboard of risk-related metrics, following political and business news and trends, responding to specific risk-related information requests from internal customers, and sending early risk warnings to those who would benefit from that information. This staff would also help local units develop their risk management capabilities. This room would also allow users to flag supplier names or purchase commodities for alerts when noteworthy information becomes available. A central role of the war room is to act as a repository for storing, interpreting, and disseminating as needed, risk-related intelligence.

Several trends make the war room an attractive possibility when thinking about supply chain risk. Without question we are seeing a general movement (at least in some areas) toward greater centralization and centrally led leadership within supply chain management. It is time for risk management to join the list of activities that would benefit from a central leadership focus. Secondly, widely dispersed supply chains and economic uncertainty are combining to make risk more rather than less critical as a supply chain topic. And finally, technology is available that can search and capture data and information in real time around the world.

Part of the war room can be outsourced to risk event specialists. One such specialist is NC4, a California company that provides detailed and real-time threat intelligence to clients about global risks and changing conditions that can affect operations and the safety of traveling employees.[19] The company scans databases and news feeds around the globe 24/7 to provide a real-time assessment of global incidents and assigns a magnitude or priority on a global situation map (along with other risk-related services). This information is available graphically at the company's command centers as well as each client's headquarters. The combination of in-depth global risk analysis, tracking of employee worldwide travels, and real-time incident alerts can provide predictive intelligence to those responsible for the well-being of employees and company assets.

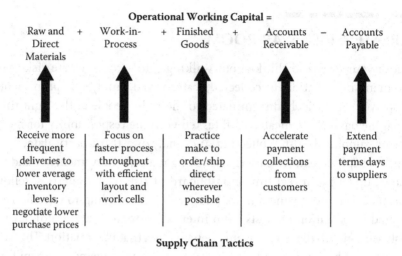

FIGURE 12.2
Managing operational working capital.

MANAGE WORKING CAPITAL

For many companies the notion of managing working capital is evolving. From an accounting perspective, working capital is simply the difference between current assets and current liabilities, information that is readily available on a company's balance sheet. Unfortunately, this is not the best way to look at working capital from a supply chain risk perspective. A different way to think about working capital is from an operational or supply chain perspective, which Figure 12.2 illustrates.

Extending payment terms to suppliers is an enticing option for managing working capital, particularly to the finance group. Unfortunately, stretching supplier payment terms directly affects, and not in a good way, relationships with suppliers. The CEO of Federal-Mogul Corporation, a tier-one automotive supplier, announced his company no longer planned on extending longer payment terms to customers after it saw a large cash outflow due to extended terms. According to the CEO, "If we need to lose market share because of our terms, I'm willing to concede business if we cannot continue to operate on the margins or the terms that are reasonable for our organization."[20] What are some of the potential risks associated with extending payments to suppliers? Stretching payment terms can

- Force suppliers to recoup extra costs by increasing purchase prices
- Affect the buyer–seller relationship to the point where suppliers do not view the buyer as a desirable customer
- Adversely affect a supplier's financial position, which could lead to supply disruptions
- Force suppliers to secure short-term financing to cover cash shortfalls
- Create unhealthy power struggles between suppliers and customers

Dozens of ways exist besides extending payment terms to better manage working capital, particularly the inventory portion of the equation. Do not forget that inventory is a complex topic that can be managed through three separate approaches—the volume, the velocity, and the value of inventory.

If your company could only select two approaches to focus its working capital improvement efforts, controlling inventory through perfect record integrity and effective demand estimation and management should be at the top of the list.

Controlling Inventory through Perfect Record Integrity

It is difficult to manage something from a risk perspective when you cannot control it. This includes the amount of capital tied up in supply chain inventory. A key objective across every part of the supply chain is to have record integrity that is as close to perfect as possible. Inventory control means that we know exactly what we have, how much we have, and precisely where it is located. Whether we have the right quantity and type of inventory and if it is positioned correctly are not answered by effective inventory control. Those topics are part of inventory management.

Record integrity or accuracy exists when the physical inventory on hand equals the computerized or electronic record on hand (POH = ROH), regardless of the quantity of inventory. Any difference between POH and ROH represents error that increases supply chain risk. This error can be the result of operationally mismanaging inventory, which affects the physical (POH) side of record integrity, or from systems-related discrepancies, which affect the computerized (ROH) side of record integrity.

When record integrity is lacking, steps must be taken to identify the sources of error with corrective action taken. In many ways the pursuit of perfect record integrity is similar to the pursuit of perfect product quality. When a nonconformance occurs, in this case a mismatch between physical inventory and its corresponding inventory record, the search for root causes should commence. This search will require asking and answering some important questions that address both the physical and informational side of inventory control, including the following:

- Are record errors displaying a random or systematic pattern across stock-keeping units (SKUs)?
- How severe are the differences between physical stock quantities and electronic quantities?
- Are proper receiving, stock-keeping, and withdrawal procedures and systems in place?
- Are suppliers shipping quantities that match their documentation?
- Are effective and timely cycle-counting procedures used?
- Is inventory scrap and obsolescence accounted for correctly?
- Is inventory theft a problem?
- Are employees trained to properly move, handle, and disburse material?

Perfect or near-perfect record integrity is never an accident. It is the result of various activities and procedures designed to ensure the amount of physical inventory equals the computerized record of inventory on hand. Poor record integrity contributes directly to greater operational and financial risk. Pursuing perfect record integrity directly reduces exposure to those risk categories.

Effective Demand Estimation and Management

Perhaps the most important information that flows across a supply chain is all the claims on a company's output for a particular period, including forecasted demand, actual orders for which commitments are already made, spare parts requirements to support aftermarket needs, and adjustments resulting from changes in inventory policies. These combined claims represent a company's demand estimate. It is startling how many systems use demand estimates as their primary data input. One could argue persuasively that demand planning is a company's most important process, particularly when managing supply chain risk.

The counterpart to demand planning is supply planning, which involves the steps taken to ensure that materials, components, and services are available to support the demand plan. Unfortunately, the coordination and hand-off of information between the demand and supply sides of the value chain, not to mention the trust between these two sides, has historically been less than stellar. While most companies engage in some form of demand planning, the link to the supply side is often incomplete.

Best-practice companies demonstrate certain characteristics related to effective demand and supply planning. The first characteristic involves the use of systems and practices that help balance demand and supply across a supply chain. Recall from an earlier chapter that two of the more advanced systems to collaboratively align supply and demand are S&OP and collaborative planning, forecasting, and replenishment (CPFR). Chapter 3 discussed S&OP and CPFR. A second characteristic is the relentless pursuit of improved forecasts and forecasting techniques. While many practices are associated with companies that excel at forecasting, the following are usually part of their portfolio.

Perhaps first and foremost, best-practice companies assign clear accountability of forecasting success to an executive or executive steering committee. This includes accountability for forecast accuracy as well as for continuously improving the forecasting system. Forecast leaders also regularly measure forecast accuracy across their different products. Forecast accuracy, which is an important supply chain metric, should be computed regularly and compared against preestablished benchmarks. Forecasting leaders also rely on quantitative systems to identify the forecasting algorithm and model that best fits their historical demand pattern, and these forecasts are updated frequently as new information becomes available.

A major part of continuous improvement involves regular reviews to identify the root causes of forecast error, similar to the discussion of inventory record errors. Review teams should determine if errors are randomly distributed, if forecasts are consistently too high or low, or if other techniques might produce better forecasts. Best practice companies apply Six Sigma and other quality improvement techniques when studying forecast error. Instead of improving the quality of a tangible product, they improve the quality of information.

Astute supply chain planners also recognize that different forecasting models fit different needs. A "one size fits all" approach is usually not the best way to forecast demand.

A final characteristic of demand estimation leaders is they do not simply react to changes in demand patterns—they try to influence these patterns by managing demand. From a risk perspective this means a shift from reacting to orders to influencing when orders occur or when they will be scheduled.

CONCLUDING THOUGHTS

This chapter presented an array of emerging risk management approaches. One thing we know for certain: There will always be a growing set of risk management tools, techniques, and approaches to either replace or enhance the existing set. We are confident about this for a variety of reasons— creative thinking and innovation is a never-ending process, continuous improvement has become a way of life, supply chain risk management is still an immature discipline, and supply chain risk is not going away anytime soon. Someone will always come out with the next great idea.

Summary of Key Chapter Points

- Because the link between positive relationships with suppliers and improved corporate performance is a strengthening rather than weakening one, becoming a preferred customer to suppliers offers advantages that could lead to future competitive advantage and reduced risk.
- A clear link exists between a customer's (i.e., the buyer) behavior, the satisfaction level a supplier has with a particular customer, and a supplier's willingness to provide preferential treatment.
- The use of an enterprise-wide, end-to-end high-level supply chain risk awareness heat map may be a good starting point for a dialogue on risk within a company's supply chain.
- No standard nomenclature specifies what a supply chain map should contain or look like. This is in contrast to project management or lean management, where the tools and usage protocols are well established.
- The growth of supply chain mapping as a risk management technique faces various challenges, particularly when a company has a large supply base or complex product structures.
- While industry and economic clusters have many redeeming qualities, less mentioned is their potential dark side. From a risk management perspective it may make sense to analyze and decluster a company's clusters.
- Even though flexibility, which is a powerful way to address supply chain risk, has been discussed for some time, most perspectives view

this concept in terms of adjusting volumes in a manufacturing environment. Supply chain flexibility involves much more than this limited perspective.

- Several trends make the risk war room an attractive possibility when thinking about supply chain risk, including greater centrally led leadership within supply chain management, widely dispersed supply chains and economic uncertainty that combine to make risk more critical, and technology that can search and capture data and information in real time from around the world.

- Dozens of ways exist to better manage working capital, including combining purchase volumes for per unit price discounts (lower the value of inventory), more frequent deliveries from suppliers (lower the average volume of inventory), and redesigning physical processes for faster throughput of material (increase the velocity of the inventory). Working capital management is becoming a primary part of financial risk management.

ENDNOTES

1. Trent, Robert. "The Supplier Satisfaction Research Project." Lehigh University, Bethlehem, PA, 2012.
2. Gilmore, Dan. "Mapping and Modeling Your Supply Chain." Accessed from http://www.scdigest.com/assets/FIRSTTHOUGHTS/13-01-15.php?cid=6737, retrieved May 24, 2013.
3. DeWitt, Tom, and Larry C. Giunipero. "Cluster and Supply Chain Management: The Amish Experience." *International Journal of Physical Distribution & Logistics Management*, 36, 4 (2006): 289–308.
4. DeWitt, Tom, and Larry C. Giunipero. "Clusters and the New Economics of Competition." (p. 291 citing Michael Porter), *Harvard Business Review* (Nov/Dec 1998): 77–90.
5. McCrea, Bridget. "Best Practices in Global Transporation and Logistics." *Supply Chain Management Review*, (December 2012): 8.
6. Lee, Heesu. "Perfecting the Face-Lift, Gangnam Style." *Bloomberg Business Week*, October 14–20, 2013: 19.
7. Chu, Kathy. "After Floods, Business Still Wary of Thailand." *The Wall Street Journal*, October 6–7, 2012: B1.
8. Sun, Albert, and Glen Goldbach. "How a Flexible Supply Chain Delivers Value." *Industry Week*, April 5, 2013, Accessed from http://www.industryweek.com/procurement/how-flexible-supply-chain-delivers-value.
9. Accessed from http://www.businessdictionary.com/definition/flexibility.html.
10. Mann, Ted. "Rail Corridor Hit with Major Outage." *The Wall Street Journal*, May 20, 2013: A3.

11. Stevenson, Mark, and Martin Spring. "Flexibility from a Supply Chain Perspective: Definition and Review." *International Journal of Operations Management*, 27, 2 (2007): 713.
12. Strumpf, Dan. "Headwinds for Rally in Natural Gas." *The Wall Street Journal*, October 8, 2012: C5.
13. Vance, Ashlee. "Stars and Stripes and Servers Forever." *Business Week*, February 28, 2011: 33.
14. Walt, V. "Meet the Third-Richest Man in the World." *Fortune*, 167, 1 (January 14, 2013): 74–79.
15. Lieber, Nick. "Suddenly, Made in the USA Looks Like a Strategy." *Business Week*, March 28, 2011: 57.
16. Simich-Lefi, David. "When One Size Does Not Fit All." *Sloan Management Review*, (Winter 2013): 15–17.
17. Lefebvre, Ben. "Trains Leave Pipeline in Lurch." *The Wall Street Journal*, May 24, 2013: B1.
18. Rondini, Denise. "Less Nickel Was Worth the Savings." *Transport Topics*, (April 29, 2013): 19.
19. Tobias, Marc Weber. "Names You Need To Know: NC4 Situational Readiness." *Forbes*. Accessed from http://www.forbes.com/sites/marcwebertobias/2011/05/01/names-you-need-to-know-nc4-situational-readiness/.
20. Ng, Serena. "Firms Pinch Payments to Suppliers." *The Wall Street Journal*, April 17, 2013: A1.

13

Risk Measurement

Imagine a measurement system that, when working effectively, offers the opportunity to reduce supply chain risk. Next, imagine the possible outcomes when such a system fails to work as intended. A number of years ago a consumer products company with $100 million in annual sales developed a scorecard system to measure supplier performance. Besides creating a system that was not validated and was less than professional in appearance, many larger suppliers challenged their scores, particularly when the scores were lower than what they received from their more sophisticated customers. The measurement system was such a nonstarter that it deterred the company from moving forward with its supplier measurement objectives. It also affected, and not in a good way, the company's relationships with its suppliers. Not much in the way of risk reduction occurred here.

Welcome to the world of measurement, a topic that can enhance or impede a company's risk management efforts. This chapter examines risk measurement from a variety of perspectives. We first discuss measurement validity and reliability, something that is critical as companies create new ways to evaluate risk. This is followed by a presentation of best-in-class supplier performance measurement systems, quantified risk indexes, and a system for measuring risk at the country level. Next, we present the increasingly talked about subject of total cost measurement. The chapter concludes with a set of emerging risk metrics.

RISK MEASUREMENT VALIDITY AND RELIABILITY

As supply chain risk management (SCRM) evolves as a discipline, it almost goes without saying that measurement will play an integral part.

Recall that measurement is one of the key risk enablers we introduced in Chapter 3. As we work with companies, we are seeing all kinds of new measures, measurement models, and risk indexes emerging that are part of the risk management process. A basic concept whenever measurement plays a central role is to ask a simple question: Is the measure or model valid and reliable?

Valid means that an indicator or model measures what it is supposed to measure. If we had to replace the word *valid* with another word, that word would be *accurate*. If a social scientist develops a scale to measure individual happiness, for example, does that scale actually measure happiness? In the risk arena, if a measure is supposed to measure the probability of a supplier failing financially, does the measure actually measure financial distress? Or, an index might translate risk scores into a system that assigns red, yellow, or green risk indicators. Is the cutoff value defining red versus yellow actually where the cutoff should be? If a supplier measure indicates a supplier is high risk, is it really a higher risk compared with other suppliers?

We do not want to give the impression that validating a measure or model is easy to do. Our concern is that far too often risk measures and indicators are developed but not sufficiently tested, usually because validation can be a time-consuming process. In the social sciences, and many observers consider business to be a social science, researchers have to address many kinds of measurement validity or risk having their work rejected by external reviewers. Different kinds of validity can include construct, convergent, face, internal, predictive, statistical conclusion, content, criterion, and concurrent validity. Validity has many dimensions, enough to give a person a serious headache.

A second important dimension of a risk measure is reliability. Reliability is the extent to which a measure provides results that are consistent from use to use. A watch could measure time (it has validity), but it could be inaccurate as its battery wears down. Or, the same piece of equipment used to measure blood pressure is not reliable if it gives different readings when no real change in a person's blood pressure occurred. Something that is reliable means that we have confidence in its use time and time again.

Possible problems with risk measures and indexes are similar to Type I and Type II measurement errors in quality management. A risk measure or index may be so sensitive that it raises a red flag when no unusual problem or risk exists (i.e., a false positive, or Type I error). After receiving enough false warnings, trust in the system erodes as users become desensitized to what the measure conveys. Another possible outcome is similar to Type II quality errors—the measure says there is no problem when in fact there is a risk event pending or likely. Unfortunately, when this is the case we are lulled into a false sense of security when the system should be picking up various signals. Perhaps the model supporting the risk measure is not sensitive enough. Or, perhaps the right factors are not part of the model, causing the model to miss some important clues.

If measurement validity is so important, how do companies ensure their risk measures are valid? Perhaps the best way to validate a risk measure or measurement approach is through simulation testing using historical data, similar to what occurs when validating forecasting models. After all, any measure that is forward looking, and most risk measures should be forward looking, is essentially a forecasting tool.

Validity and Bridge Safety Measures

Here is an example of model validity that fell out of nowhere. When an oversize truck traveling in Washington State hit a bridge girder, causing an entire section of Interstate 5 to fall into the Skagit River, it was not long before the system that calculates sufficiency ratings for bridges came under scrutiny.[1] And the verdict of this scrutiny was that the sufficiency rating system to assess bridge safety has some serious shortcomings.

Part of the problem is the complexity of the sufficiency rating system developed in the 1970s. About 20 factors, almost half of which have nothing to do with a bridge's actual condition, are put into a magic formula that generates a single sufficiency bridge rating. Mathematically, it is possible that a bridge that is more vulnerable to collapse has a higher sufficiency rating than a bridge that is less vulnerable. A sufficiency rating less than 80 is necessary to qualify for federal funding for bridge modifications, while a rating under 50 qualifies a bridge for replacement. In other words, serious decisions are made because of the sufficiency rating. And it appears that the rating system may not be doing what it is supposed to do.

Compiling so many factors into one rating increases the probability that serious deficiencies are overshadowed by other factors, such as average daily traffic and detour length if a bridge is taken out of service. Shortcomings in the current rating system are causing engineers to look at new ways to measure bridge risk, including using software to predict how bridges will change and possibly fail over time, along with cost-benefit analysis to optimize spending on maintenance and repair. However the measurement of bridge hazard risk eventually turns out, the one thing that is becoming increasingly clear is that bridge sufficiency ratings are not all that sufficient.

SUPPLIER PERFORMANCE MEASUREMENT— DOING IT RIGHT

Most firms, particularly larger ones, will say they have some sort of supplier performance measurement system in place. Many companies call the output from these measurement systems *supplier scorecards*. Our discussion here is not a how-to on supplier performance measurement; other sources have covered this topic quite well. Rather, we address the issues that tend to affect supplier measurement systems at most companies. Let's highlight these shortcomings through a case example.

The Case of the Deceptive Scorecards

During a review of a supplier scorecard system at a global logistics company, a training instructor asked a buyer to name one of his best-performing suppliers in terms of its performance score. Without hesitation the buyer provided a supplier's name. Another participant in the room responded quickly by saying that in the operations facility this is one of the worst suppliers his group deals with on a day-to-day basis. How can one person say this is a supplier worthy of a preferred status while another person would like to see this supplier go away? And, perhaps most importantly, what are the risks of a measurement system that awards high scores (and likely future business) to what may be poorly performing suppliers? The irony here is that a system that is designed to reduce supply chain risk could actually be increasing risk.

These differences of opinion resulted in a spirited discussion among the participants in the room. During this discussion the group reached consensus about a number of important points. First, the group agreed that although the measurement system is supported by an extensive database that allows all kinds of on-demand analyses, the data to support that system are largely collected and input manually. Furthermore, many performance items require subjective judgments. Second, most buyers had responsibility for inputting data quarterly for about 25 suppliers, a heavy burden that is in addition to their normal workload. Many in attendance also agreed that the data are input just before, and sometimes after, the quarterly cutoff. Third, attendees acknowledged that supplier scores are used as one indicator of a buyer's job performance, potentially creating a conflict of interest.

The group also agreed that all suppliers are essentially held to the same criteria with the same assigned weights, even though no one believes that suppliers are equally important or similar. Participants further agreed that internal customers have no way to provide input into the measurement process, even though this group has the best perspective regarding a supplier's day-to-day performance. Some participants were even confused about how to rate a supplier since some suppliers provide material from more than one site. Finally, no clear agreement emerged that the measurement process was contributing to better supplier performance. It was taken as an article of faith that measurement is a worthwhile pursuit.

Table 13.1 provides a set of guidelines for assessing whether a supplier performance measurement system is likely to satisfy its intended use. Evaluating a measurement system against these criteria will help ensure the system is leading-edge. In fact, the items that appear in this table essentially define the characteristics of a world-class supplier performance measurement system. If supplier measurement is an important objective, then let's at least do it right. Measuring performance incorrectly is an invitation to trouble, and we all know that trouble and risk are best friends.

TABLE 13.1

Characteristics of an Effective Supplier Measurement System

- The measurement system allows scoring flexibility so all performance categories and suppliers are not measured the same way.
- Internal customers evaluate supplier performance through an online portal that feeds information directly to the measurement system.
- Performance reports are forwarded electronically to suppliers with review and acknowledgment required by executive supplier management.
- Each location at a supplier receives an operational performance report while the supplier's corporate office receives a "relationship" performance report.
- Supplier performance reports include total cost measures wherever possible instead of price measures.
- Supplier performance, particularly cost, quality, and delivery, is updated in real time as transactions occur.
- The measurement system separates critical suppliers from marginally important suppliers.
- The supplier measurement database allows user flexibility when retrieving and displaying data.
- The measurement system provides early-warning performance alerts such as predicted late deliveries from suppliers.
- Suppliers have the ability to view their performance online with comparisons against other suppliers.
- The measurement system is regularly compared against best-practice companies.
- Real performance improvement can be demonstrated as a result of the measurement system.

QUANTIFIED RISK INDEXES

This approach scores risk events or suppliers using algorithms that model risk. Risk indexes are quantitative models that consider multiple factors to arrive at a single risk indicator score, similar to the bridge sufficiency rating discussed earlier. These indexes consider more than simply a supplier's financial status when arriving at a risk score. To date most risk indexes or indicators have been internally developed since third parties have been slow to respond with new tools and approaches, something that will likely change.

It is no surprise that risk indexes range from basic to complex. On the basic side, some companies use a variety of simple algorithms to arrive at a single risk score or index. A company might consider the probability of a potential risk event occurring using a 1–10 scale. That figure is then multiplied by the effect a risk event would have on a 1–10 scale. Other factors can be introduced, such as the ability to detect the risk on a scale of 1–10 (a higher score means less ability to detect the risk). The three scores would be multiplied together to arrive at an overall risk index, with higher scores representing higher risk exposure. A later example will further highlight this approach. A concern about the validity of these simple risk indexes measures is particularly relevant here.

It is not unusual to translate risk index figures into a red/yellow/green visual system (recall from Chapter 6 this method is used with the Z-Score). This should also happen during new product development as specific risks are identified, something that Chapter 4 addressed. Product development teams will work to address red and even yellow items prior to product launch. The goal should be to launch new products with as few red risks as possible. Why wait until after the fact?

Interestingly, a body of research is emerging that counters the notion that complex algorithms and models are automatically more effective than simple rules of thumb or guidelines when making organizational decisions. This is something to keep in mind when developing risk indexes. A hypothesis put forward is that complex situations create so many possible courses of action and become so complex to use that individuals become confounded, often to the point where they delay decisions, default to the safest option, or avoid making choices altogether.

Research suggests that simple rules can equal, and at times exceed, the effectiveness of more complicated analyses across a range of decision areas.[2]

While the analysis and data that lead to the rules may be sophisticated, and at times will even be complex, the rules that result should be elegant in their simplicity as they provide guidance to users. When the up-front work to develop the risk index is rigorous, the chances are good the risk index will have validity. This is something to keep in mind when the temptation exists to develop overly complex algorithms, models, and risk indexes.

A Risk Index Example

Consider the following example, which is a technique used by a food manufacturer to develop risk indexes to support its risk management efforts.[3] This company examines specific risks from three dimensions: severity of the risk, the probability of the risk occurring, and the probability of early risk detection. This approach is consistent with the FMEA (failure modes and effects analysis) approach, a widely used quality management technique that considers these three factors. Let's look at an example:

	Risk: Poor Product Freshness
Severity (1 *low*–10 *high*)	7
Probability of Occurrence (1 *low*–10 *high*)	5
Probability of Early Detection(1 *high*–10 *low*)	3
Risk Index (7×5×3)	105

Companies that use this approach identify, evaluate, and then rate all possible risks, which are then addressed in terms of priority. As mentioned, with any risk index we must be concerned about its validity. Here, what defines the incremental values in each scale (i.e., low to high is a broad range with a great deal in between)? Is the incremental difference between a score of 3 and 4 in a scale the same as between 5 and 6 or 8 and 9? Should the three categories be equally weighted, which is the case here? Would three people looking independently at the same risk arrive at the same or similar score using this tool? (This is called inter-rater reliability). Too many companies use tools such as the one presented here without fully performing the up-front work necessary to validate the tool. The virtue of this approach is its simplicity. It is easy to understand and use.

Country Risk Indexes

A variety of risk indexes are available that evaluate country risk. One of the more comprehensive of these indexes, and one that will be featured here, is the International Country Risk Guide (ICRG) composite risk rating produced by the Political Risk Services (PRS) Group.[4] This composite risk rating by country, which is updated monthly, includes 22 variables across three subcategories of country risk—political, financial, and economic. A separate detailed scoring and weighting methodology is created for each of the 22 subcategories.

The composite risk rating by country is based on 100 total points with the political risk rating comprising 50% and the financial and economic risk ratings contributing 25% each. The composite scores, ranging from zero to 100, are then broken into categories from *very low risk* (80 to 100 points) to *very high risk* (0 to 49.9 points).

At the enterprise risk management (ERM) level, country risk index ratings are valuable when a company is thinking about making foreign direct investments. At the SCRM level, this information can influence logistics, sourcing, and selling activities. Our advice is to not even bother creating country risk indexes. Third-party indexes are available that are more comprehensive than anything a company could construct acting on its own.

USING TOTAL COST MEASURES TO MANAGE RISK

Total cost is a topic that companies cannot ignore as they search for new and better ways to manage supply chain risk. What exactly is total cost? Total cost includes the expected and unexpected elements that increase the unit cost of a good, service, or piece of equipment. The logic behind the development of total cost systems is that unit cost or price never equals total cost. And, as the gap between unit cost and total cost becomes progressively larger, so does a company's risk exposure. Regardless of where a company applies total cost models, these models all attempt to capture data beyond unit price.

In uncertain times, the need to understand every element of cost has never been greater. Total cost systems help management to identify the impact of different cost elements, to track cost improvements in real terms over time, and to gain management's attention regarding where cost reduction efforts will have their greatest payback across the supply chain. While there is some overlap between sources regarding which cost elements to include in total cost models, no agreement exists regarding exactly what these models should contain. This issue becomes more complex once we understand that, like forecasting models, different types of cost models exist. And, like forecasting models, total cost models almost always have some inherent inaccuracy to them.

Types of Total Cost Models

Every total cost model is part of a family of measurement systems called cost-based systems. Like any measurement system, cost-based systems offer advantages and disadvantages. These systems can be extremely challenging to develop and use (if they were not challenging then everyone would routinely use them).[5] Some of the challenges include (1) relying on data that are derived across global supply chains, (2) maintaining the discipline across the supply chain to use the models routinely, and (3) using data that are not known with certainty or are estimates of what might happen. A mean time between failures (MTBF) estimate, for example, is often used when evaluating the total expected cost of capital equipment. How close is the estimate to what actually happens in terms of equipment reliability?

Across a supply chain we generally see three kinds of total cost models in use, regardless of whether this involves a domestic or international supply chain. These include total landed cost models, supplier performance index models, and life-cycle cost models.

Total Landed Cost Model. A total landed cost model is used when evaluating suppliers prior to making purchase decisions, although that is not the only time when a landed cost model should be used. Total landed cost is the sum of all costs associated with obtaining a product, including acquisition planning; unit price; inbound cost of freight, duty, and taxes; inspection; and material handling for storage and retrieval.[6] Each of these cost categories will also contain numerous subcategories. Best-practice companies require their commodity teams or buyers to attach spreadsheets that show the total landed cost of a purchase requirement whenever they propose a supply strategy or make a supplier selection decision.

Total landed cost models should also be used when doing business with suppliers on an ongoing basis. The factors that affect the sourcing decision in the first place are dynamic and subject to change (think transportation and exchange rates here). Furthermore, actual costs should replace any estimated or forecasted costs in the model as they become available. This helps to validate the assumptions in the model.

When developing total landed cost models it is best to start with the unit price and then build up the total cost as goods move from origin to destination. Ideally every cost element is presented in the same unit of measure. If a product is priced by the pound, then every corresponding cost element in the model should appear as a cost per pound. The cost elements in landed cost models should be divided into categories that reflect a logical progression through the supply chain:

- Unit price—unit price usually appears on the first line of the cost model
- Within country of manufacture costs—includes materials, storage, labor, quality, overhead, obsolescence, packaging, risk or disruption, exchange rates, inventory carrying charges
- In-transit to country of sale costs—includes transportation charges, fuel surcharges, insurance, port charges, handling, security, banking fees, broker fees, potential detention charges, duties, handling agency charges, inventory carrying charges
- Within country of sale costs—includes local transportation and handling, storage fees, taxes, safety stock, inventory carrying charges, yield, productivity implications, maintenance, quality, overhead allocation, payment terms

Supplier Performance Index (SPI) Model. Various models attempt to capture the true cost of doing business with a supplier on a continuous basis. Perhaps the best known of these models is something called the Supplier Performance Index (SPI). SPI calculations, which focus largely on supplier nonconformance costs, are helpful when tracking supplier improvement over time, quantifying the severity of performance problems, deciding which suppliers to eliminate from a supply base, and when establishing minimum acceptable levels of supplier performance.

The SPI is a total cost model that presents its output in the form of an index or ratio. It assumes that any quality or other infraction committed by a supplier during the course of business increases the total cost (and hence the total cost performance ratio) of doing business with that supplier. This approach is more applicable after supplier selection because it is populated with cost occurrences that have happened rather than are expected to happen. The SPI calculation for a specific period is a straightforward formula:

$$\text{SPI} = (\text{Cost of material} + \text{Nonconformance costs})/(\text{Cost of material})$$

Assume a supplier delivers $1 million worth of parts to a company in the third quarter of a year. The supplier also commits three infractions that quarter—a late delivery, missing documentation, and a parts shortage. Further assume the buyer assigns $30,000 in total nonconformance charges to these infractions. The supplier's SPI for the third quarter is 1.03, or (($1,000,000 + $30,000)/$1,000,000). The SPI of 1.03 means the total cost of doing business with this supplier is 3% higher than the unit price. If the unit price of a supplier's good is $127, then the estimated total cost of that item is really $130.81 ($127 × 1.03). Because the SPI is a standardized metric, it allows comparisons between suppliers. A supplier with a higher SPI has a higher total cost than one with a lower SPI. It is important to compare suppliers within the same commodity to ensure "apples to apples" comparisons.

SPI Drawbacks. Although the SPI can be an effective tool, it is by no means perfect. In fact, it has some potential drawbacks that users must understand. First, because it is an index, the actual unit cost of an item from a supplier is not considered directly in the SPI calculation—only the value of the total shipments and infractions are considered. A higher supplier unit cost inflates the value of the shipments compared with a supplier that has a lower unit cost, making any infractions look smaller given the shipment value. Mathematically, this makes the SPI value lower, all else equal, for the higher-price supplier. Let's illustrate this with two suppliers that ship the same number of units with the same infraction charges but with different unit costs:

Supplier A	Supplier B
50,000 units @ $9.00 per unit	50,000 units @ $10.50 per unit
$27,500 nonconformance charges	$27,500 nonconformance charges
SPI = ((50,000 × $9) + $27,500)/$450,000	SPI = ((50,000 × $10.50) + $27,500)/$525,000
= 1.06	= 1.05

The difference here is a mathematical artifact of the different unit costs. Ideally, the buying company would employ a total landed cost model during supplier selection so any issues regarding differences in unit costs would have already been considered.

The SPI calculation also has a built-in bias against small volume suppliers. Assume three suppliers within a commodity group commit the same infraction that resulted in a $3,000 nonconformance charge. The first supplier provided $15,000 worth of goods during a quarter, the second supplier provided $10,000 worth of goods, and a third supplier provided $30,000 worth of goods. The SPI for the first supplier is 1.20 (($15,000 + $3000)/$15,000), the SPI for the second supplier is 1.30 (($10,000 + $3,000)/$10,000), and the SPI for the third supplier is 1.10 (($30,000 + $3,000)/$30,000). Even though each supplier committed the same infraction, the smaller supplier appears worse from an SPI perspective, particularly compared with the larger supplier.

This bias requires the calculation of a Q adjustment factor, which is essentially a weight applied to the nonconformance costs. The adjustment factor allows valid SPI comparisons by removing the bias against suppliers with a lower total value of deliveries. It makes sense to calculate an adjustment factor that removes this bias if suppliers within a commodity provide widely differing volumes. If we want to make total cost models as accurate as possible, then we have to think about the Q adjustment factor. Figure 13.1 illustrates the step-by-step calculation of the Q adjustment factor.

A final drawback to the SPI approach is that it requires a great deal of discipline and cross-functional support to stay on top of the required data collection. Supplier infractions can occur or be discovered at different points along a supply chain. Inconsistency in collecting and allocating supplier charges will quickly undermine the validity of this model.

	Supplier A	Supplier B	Supplier C
1st quarter deliveries	10	12	10
Total value of deliveries	$10,500	$18,000	$35,000
Average delivery value	($10,500/10) = $1,050	($18,000/12) = $1,500	($35,000/10) = $3,500
Non-conformance charges assigned to each supplier	$1,000	$1,400	$2,500
1st quarter SPI	($10,500+$1,000)/$10,500 = 1.10	($18,000+$1,400)/$18,000 = 1.08	($35,000+$2,500)/$35,000 1.07
Average shipment from all suppliers*	$2,500	$2,500	$2,500
Q adjustment factor	$1,050/$2,500 = .42	$1,500/$2,500 = .60	$3,500/$2,500 = 1.4
Adjusted SPI	1.04	1.05	1.10

Adjusted SPI for Supplier A = $10,500 + ($1,000 × .42)/$10,500 = 1.04
Adjusted SPI for Supplier B = $18,000 + ($1,400 × .60)/$18,000 = 1.05
Adjusted SPI for Supplier C = $35,000 + ($2,500 × 1.4)/$35,000 = 1.10

* Average shipment from all suppliers equals (total value of shipments/total shipments) for all suppliers within a commodity, not just those listed in this table. This is a provided piece of data in this example.

FIGURE 13.1
Supplier performance index with Q adjustment.

Life-Cycle Cost Models. Life-cycle cost models may be what comes most to mind when thinking about total costs analysis. This type of model is most often used when evaluating capital decisions that cover an extended time period, such as equipment and facilities. Life-cycle models are very similar to net present value models used in finance. Life-cycle cost models are used when evaluating capital decisions, such as plant and equipment, rather than the purchase of everyday components and services. The other cost models described here are more applicable for repetitively purchased goods or services. Life-cycle costs apply whether equipment is sourced domestically or internationally.

Developers of life-cycle cost models often allocate their cost elements across four broad categories that reflect usage over time. The life-cycle is essentially one of buying, shipping, installing, using, maintaining, and disposing:

- Unit Price—Includes the price paid along with purchase terms
- Acquisition Costs—Includes all costs associated with delivering equipment, such as buying, ordering, and freight charges to the customer
- Usage Costs—Includes all the costs to operate the equipment, including installation, energy consumption, maintenance, reliability, spare parts, and yield and efficiency during production
- End-of-Life Costs—Includes all costs incurred when removing equipment from service, less any proceeds received for resale, scrap, or salvage

Companies should compare the assumptions made during the development of life-cycle estimates with actual data as they become available, particularly since life-cycle models often look years into the future. This will provide insights regarding how to improve the life-cycle models.

A popular misconception is that having a total cost model inherently provides better information than not having a total cost model. We like to think this is true, but the reality is that total cost models, like forecasting models, almost always have some degree of inaccuracy. This is especially true if a model is populated with data that are based largely on estimates or averages rather than actual data. Or, the model may fail to take into account some important cost elements. Do not underestimate the value, however, of a well-specified total cost model when managing supply chain risk. Total cost measurement is one of the best risk management approaches that we can put in place today.

SUPPLIER CAPACITY ESTIMATE MEASURES

Most of us have heard stories about a buyer placing an order with a supplier only to find that the supplier's assurances about available capacity simply are not true. Or, how about when total demand in a marketplace increases and a buying company finds it has been placed on the short end of a supplier's allocation schedule. Either of these conditions creates operational risk as supply is not readily available to satisfy demand.

A risk management approach that will provide some clues into available capacity is something we call *rough cut supplier capacity analysis*. It is called *rough cut* because it is not meant to be a precise estimate of available supplier capacity. We are simply trying to get a feel regarding what might be available and comparing that to a specific requirement. And when we get a better feel for the amount of available capacity, we can engage in some worthwhile discussions with suppliers. Let's illustrate how this technique works.

Table 13.2 presents data for the three suppliers that appeared in Chapter 6 when calculating Z–Scores. We are now introducing two new additional pieces of data—the estimate of the average capacity utilization rate at each supplier and a quoted price per unit for the item of interest. This table presents a methodology for estimating the capacity available at each supplier. This analysis reveals that only one supplier appears to have adequate capacity available. According to these numbers the buying company could face major operational risk if it decided on a single-source contract with FASE Chemicals. The risk could be so severe that it affects the success of a product, which then leads to strategic risk. And placing a single-source contract with DMS will require some serious discussions to verify these estimates and, if verified, to work out a plan to meet demand requirements.

The point of this exercise is to provide a broad understanding of where a supplier stands in terms of capacity. One risk here is that a supplier could also be speaking to other buyers that are interested in the available capacity. It is also possible some contracts at the supplier are expiring, which could make additional capacity available. One thing we are confident about is these suppliers are not likely to drop existing customers simply to fulfill a new contract, at least in the short term, making the capacity discussion a key part of the risk assessment process.

TABLE 13.2

Rough Cut Supplier Capacity Analysis

A pharmaceutical company has forecasted that it requires 20 million pounds of a chemical compound to support the launch of a new product next year. The following data are collected to help in the estimate.

	Ninaka Materials	FASE Chemicals	DMS NV
Quoted price per pound	$4.75	$5.75	$5.20
Current installed capacity utilization	98%	95%	94%
Sales 2014 generated from the chemical compound	$6,500,000,000	$550,000,000	$1,355,000,000
Rough cut estimate of available capacity	27,926,960 lbs.	5,034,325 lbs.	16,632,570

Current installed capacity utilization indicates that portion of the supplier's production capacity that is currently utilized for the production of chemicals. For example, if current installed capacity is 98%, then this supplier is utilizing 98% of its production capacity and therefore has 2% of its capacity available for new business. This does not indicate how many available pounds this represents.

Ninaka

$6,500,000,000 sales generated from the compound/98 capacity points used to generate the sales = $66,326,530 in sales generated by each point of used capacity × 2 capacity points available = $132,653,060 potential capacity available in dollars/$4.75 quoted price per pound =

27,926,960 estimated pounds available

FASE

$550,000,000 sales generated from the compound/95 capacity points used to generate the sales = $5,789,474 in sales generated by each point of used capacity × 5 capacity points available = $28,947,368 potential capacity available in dollars/$5.75 quoted price per pound =

5,034,325 estimated pounds available

DMS

$1,355,000,000 sales generated from the compound/94 capacity points used to generate the sales = $14,414,894 in sales generated by each point of used capacity × 6 capacity points available = $86,489,362 potential capacity available in dollars/$5.20 quoted price per pound =

16,632,570 estimated pounds available

EMERGING SUPPLY CHAIN RISK METRICS

We are witnessing a combination of existing measures as well as the development of entirely new measures being applied to the risk management arena. The following describes some of these risk-related measures.

Value at Risk

A metric that will increasingly have corporate-level visibility as a risk metric is value at risk (VaR), a metric that is used extensively by the financial community to evaluate financial investments. From a financial perspective, VaR represents the largest loss likely to be suffered on a portfolio position over a holding period (usually 1–10 days) with a given probability (confidence level). VaR is a measure of market risk and includes three components—a time period, a confidence level or percentage, and a loss amount or percentage attached to a risk.[7] This concept is now being applied to supply chain scenarios. Chapter 5 provided a detailed illustration of this metric and its use in supply chain risk management.

Time-to-Recovery

The time-to-recovery (T-t-R) from adverse events, a risk resiliency measure used extensively in the information technology arena, is generating significant interest as an evolving supply chain risk measure. An example of a T-t-R measure involves the time to recover from a natural disaster, such as a flood. Virtually any supply chain disruption can have a T-t-R measure attached to it.

T-t-R measures, like most measures, have an objective attached to them. Some sources refer to this objective as the recovery time objective (RTO), which is the time in which a system, a facility, or equipment must be restored after a disruption to avoid an unacceptable break in business continuity and the incurrence of significant losses. T-t-R measures how long it takes an entity in a supply chain to reach full volume or full operating status after a major disruption. This does not have to mean full recovery at a specific facility. If the full volume for a part from a supplier can be provided within two weeks by working overtime at another facility, then two weeks is the T-t-R, even if the affected facility takes longer to recover.

Risk Exposure Index

An extension of the T-t-R index is the Risk Exposure Index, developed by David Simchi-Levi at MIT. This index attempts to provide a better way for companies to quantify their supply chain risks compared with the traditional 2 × 2 matrix that places the likelihood of an event occurring (high or low) against its financial impact (low or high). With this traditional approach, potential risk events are plotted into one of four quadrants with those with the highest likelihood of occurring and the largest impact receiving priority for attention. While this approach serves a worthwhile purpose, it is does not represent cutting-edge risk measurement.

The Risk Exposure Index assigns a cost that would occur from a potential disruption across each level and node in a given supply chain, based on the T-t-R for each level/node and the resulting financial impact (FI), including market share losses. Those individual risk components are then totaled to produce a full FI for the entire supply chain.[8]

This methodology also addresses unpredictable risks, including natural disasters and fires at critical suppliers. While these types of risks are nearly impossible to predict, the chances that such a disruption will occur over a period of time is quite likely, which is built into the model. The key to this analysis is the calculation of the financial damage that such a disruption would likely cause.

Supply Chain Key Performance Indicators

We expect to see various key performance indicators (KPIs) emerge that are viewed by executive management as supporting a company's risk management efforts. The following presents six measures, some of which have been in existence for a while but not necessarily visible at higher executive levels. These are not predictive measures—they do not provide warning about pending or specific risk events. Rather, they provide insight into how well some important business processes are operating. We will probably all agree that when important processes are operating well, a business faces lower risk exposure.

Forecast Accuracy. As mentioned in Chapter 12, best-practice companies track and assign clear accountability for forecasting success to an executive or executive steering committee. Best-practice companies also regularly measure forecast accuracy across their different products.

One survey revealed that fully two-thirds of respondents reported forecast accuracy between 50% and 80%, a range that indicates room for improvement. Forecast accuracy should be computed regularly and compared against preestablished benchmarks. Many techniques are available for assessing forecast accuracy, including mean forecast error, bias, mean absolute deviation, mean absolute percent error and tracking signals. All of these techniques compare, in some manner, actual demand against forecasted demand with the difference between these two figures considered error.

Concept-to-Customer Cycle Time. Product development leaders rely on an important time-based metric called concept-to-customer (C-to-C) cycle time. This metric reflects the importance of being aware of the time it takes to develop new products as well as acting as a target that no single functional group can unilaterally attain. Surprisingly, in our experience most companies do not measure an overall cycle time, making the development of a C-to-C cycle time measure especially attractive. Holding functional groups mutually accountable for this measure sends a powerful message about the importance of collaborating during product development efforts. And as Chapter 4 pointed out, the linkage between strategic risk exposure and new product development success (or failure) is strong.

Inventory Accuracy. Recall from Chapter 12 that measuring inventory accuracy is essential for managing various kinds of supply chain risk. Inventory accuracy exists when the physical inventory on hand for an item equals the computerized or electronic record on hand (POH = ROH), regardless of the quantity of inventory. Supply chain managers should become almost evangelical in their quest for perfect record integrity, including the integrity of records and data at suppliers and distributors.

Order-to-Cash Cycle Time. An important part of any supply chain is the customer order fulfillment process. The order-to-cash cycle involves the steps from acquisition of a customer's order to receiving payment from a customer. When viewed narrowly, order fulfillment focuses mainly on the acts of distribution and logistics. When viewed broadly, order fulfillment includes all the steps and activities from the sales inquiry to delivery, and perhaps even the return of the final product or service. This involves order preparation, transmission, entry, order filling (which may include production and purchasing), billing, shipping, tracking, and returns.

Companies that take a broader view of order fulfillment extend their perspective to include the management of accounts receivable, making order-to-cash cycle time a key performance indicator. This stresses the financial aspects of the fulfillment process by not viewing the process as complete until customer payment is received. Ineffective management of the order-to-cash cycle time has clear financial risk implications.

Perfect Order Rate. The perfect order metric is a mathematic composite of multiple factors. The perfect order is one that is delivered on time, complete (all ordered items are in the shipment), damage free, accurate (correct items and quantities), with proper documentation. Perfect orders not only drive customer loyalty for the product and producer, but they also lead to greater supply chain efficiency and reduced investments in inventory.[9]

To date the primary users of the perfect order metric have been consumer packaged goods companies. Virtually any company, however, can use a perfect order measure if they can overcome the humbling nature of this metric. As variables with a less-than-perfect value are combined, the resulting metric becomes lower and lower. This is clear from Figure 13.2. To some, it simply looks better to report each line item separately and forego the cumulative measure. Let's not mess up some good numbers with facts.

Return on Assets. If there is one higher-level measure that tells how well a supply chain is performing, return on assets is that measure. While multiple versions of this metric exist, they all include a numerator that includes income and a denominator that looks at assets. Regardless of the specific formula used, one thing we know for certain is that any supply chain problems, including the consequences of any risk events, will show up in the numerator and/or denominator of this metric. Chapter 3 illustrated how a company uses return on assets as its primary way to measure the performance of its business units.

A perfect order is an order that is delivered complete, on time, in perfect condition, and with accurate and complete documentation.

- "Perfect orders":
  - Orders delivered on time ······························► **97%**
  - Orders delivered complete ······················► **98%**
  - Orders delivered damage free ···············► **99%**
  - Orders filled accurately ·····························► **96%**
  - Orders billed accurately ···························► **97%**

Perfect Order Rate = 0.97 × 0.98 × 0.99 × 0.96 × 0.97 = 87.6%

FIGURE 13.2
Perfect order measure.

CONCLUDING THOUGHTS

Supply chain managers must take an unbiased view of their performance measurement systems. The objective here should be to take poor measurement systems and make them better while transforming good systems into excellent ones. One course of action is to assemble internally a team to compare the current state of supply chain risk measurement areas against an ideal future state. Any gaps that exist between the current and future state require a clear plan to get to a preferred state.

While measuring various dimensions of supply chain risk is a worthy pursuit, the reality is that risk measurement is simply an activity. Activity means nothing unless it leads to lower supply chain risk compared with what would likely occur without the measurement system in place. Measurement activity must lead to accomplishment.

Summary of Key Points

- A central question when thinking about supply chain risk measurement is whether a measure, model, or index is valid and reliable.
- Most firms have some sort of supplier performance measurement system in place. Fortunately, a set of best-practice guidelines exists for assessing whether a supplier performance measurement system is likely to satisfy its intended use.
- One type of measure scores risk events or suppliers using derived algorithms that model risk. Risk indexes are quantitative models that consider multiple factors to arrive at a single risk indicator or score.
- A body of research is emerging that counters the notion that complex algorithms and models are automatically more effective than simple rules of thumb or guidelines when making organizational decisions.
- At the ERM level, country risk index ratings are particularly valuable when thinking about making foreign direct investments. At the SCRM level, this information can influence logistics, sourcing, and selling behavior.
- Total cost of ownership is a topic companies cannot ignore as they search for new and better ways to manage supply chain risk. We expect to see at least three major supply chain cost models in use: total landed cost models, supplier performance cost models, and life-cycle cost models.

- A variety of existing and new measures are emerging to help manage supply chain risk, including value-at-risk, time-to-recovery, and risk exposure indexes.
- Various KPIs will emerge that support a company's risk management efforts, including forecast accuracy, inventory accuracy, the perfect order, concept-to-customer cycle time, order-to-cash cycle time, and return on assets. An underlying commonality linking these measures is that they are superordinate, which means they are higher in class, rank, or status compared with other measures.

ENDNOTES

1. Bialik, Carl. "Will This Bridge Fall? It's Hard to Say." *The Wall Street Journal*, June 1–2, 2013: A2.
2. Sull, D., and K. Eisenhardt. "Simple Rules for a Complex World." *Harvard Business Review*, 90, 9 (2012): 69–74.
3. Adapted from Dittmann, J. Paul. "Managing Risks in the Global Supply Chain." www.scmr.com/article/managing_risks_in_the_global_supply_chain
4. Accessed from http://www.prsgroup.com/ICRG_Methodology.aspx.
5. For a PowerPoint presentation of total cost system, send a request to rjt2@lehigh.edu.
6. Cowman, K. "Material Costs." *Materials Management and Distribution,* 49, 7 (September 2004): 73.
7. Accessed from http://www.businessdictionary.com/definition/value-at-risk-VAR.html.
8. Accessed from http://www.prweb.com/releases/2012/3/prweb9259939.htm, March 8, 2012.
9. Shaw, Tim. "In Search of the Perfect Order." Accessed from https://www.teradata.com/article.aspx?id=1646.

14

Learning from Risk Management Leaders

Although most companies have a way to go before they can brag about their risk management prowess, some companies have achieved distinction, at least in some aspect of risk management. This chapter highlights a variety of companies that have demonstrated their commitment and skills when pursuing supply chain risk management (SCRM): Boston Scientific, Boeing, IBM, Cisco, Delphi, and a major defense contractor. The purpose of this chapter is to appreciate what leading companies are doing to become risk management leaders. We also highlight a company that offers risk-related lessons learned the hard way.

MAKING RISK MANAGEMENT A PRIORITY AT BOSTON SCIENTIFIC

A company that is widely recognized at being at the top of its risk management game is Boston Scientific Corporation (BSC), a company started in 1979 with 38 employees and $2 million in sales.[1] Today, with a worldwide workforce of 24,000 employees, more than $7 billion in sales from more than 100 countries, and a product portfolio containing 15,000 products, it is not surprising that a company this complex continuously faces uncertainty.

Even before the 2008 economic downturn BSC had taken a heightened interest in the impact of supplier risk on the company's operations. In this regard the company is an early risk management adopter. The company created a detailed Supplier Risk Management program to help prepare for any anticipated and unanticipated risks they may face. BSC defines supplier risk management as a proactive and systematic process for cost-effectively identifying and reducing the frequency and severity of unwanted events in the supply chain that have an adverse effect on the business.

The primary goal of BSC's program is to move from being a reactive risk taker to being proactive toward risk, thereby allowing the company to reduce its overall risk exposure. The company divides this goal into four specific objectives—gain visibility to high-risk suppliers that require attention, identify and understand the specific drivers that increase supplier risk, proactively manage and mitigate supply chain risk, and measure risk mitigation and its impact.

BSC followed a three-step process when designing its risk management program. The first step, information acquisition, required BSC to gather and access information from the external environment, suppliers, and the analysis of different parts and components. The second step involved compiling this information using basic risk management systems and tools. The final step involved communicating this information to different Boston Scientific business units and plants as well as to suppliers.

Having the Right Tools

After its initial analysis, Boston Scientific designed a formalized process to manage supplier risk. This process includes (1) identifying risk areas, (2) analyzing and prioritizing these risks, (3) developing risk mitigation plans to address high-risk areas, and (4) tracking high-risk areas.

To support this process the company has developed a primary risk tool it calls the Supplier Risk Wheel. Each supplier has its own wheel. The purpose of this tool is to identify high-risk events and risk categories that require action. The Supplier Risk Wheel starts with data and survey inputs to identify specific risk events, which create the outer ring of the wheel. Risk categories that contain these identified risk events comprise the middle circle of the wheel. Each risk item (outer ring) and risk category (middle ring) is assigned a color based on the level of risk it involves (red = *very high risk*, light red = *high risk*, yellow = *medium risk*, light green = *low risk*, dark green = *very low risk*), and each level affects the next. Finally, the risk categories are used to calculate the overall supplier Risk Probability Index (RPI), which is the center circle of the wheel.

BSC has developed other tools, including a Risk Distribution Matrix and a Supplier Comparison Report, to provide further risk insight. The Risk

Distribution Matrix is a mapping tool with an x- and y-axis. The x-axis represents the supplier's Risk Probability Index from the Risk Wheel and the y-axis represents the potential revenue impact presented by the supplier. The matrix uses red, yellow, and green zone designations.

A Supplier Comparison Report allows the comparison of the highest and lowest indicators or ratings in various areas between suppliers. This approach also relies on a color-coded scheme and considers factors such as the supplier's percentage of revenue from medical devices, quality, alignment with BSC, accreditation, delivery, capacity utilization, plant size, and service support. Another tool is a comprehensive risk template that details the kinds of actions taken to prevent or mitigate risk in the case of a risk event. Boston Scientific uses all of this information to determine the best ways to eliminate or manage supplier risk.

Boston Scientific has also created a Risk Alert and Communications System. This system relies on a variety of data sources in four categories (financial, governmental, disasters, and market dynamics) to gain insight into potential risk events. Examples of data sources supporting the alert system include Dun & Bradstreet information, U.S. Environmental Protection Agency (EPA) e-newsletter, regulatory updates, U.S. Food and Drug Administration (FDA) updates, weather updates, commodity analyses, supplier communications, foreign travel warnings, and state department fact sheets. Finally, the company actively benchmarks and compares its risk management capabilities against other best-in-class companies.

Is all this work worth it to Boston Scientific? The company has developed three sophisticated metrics for evaluating its risk standing—cross product, which is a measure of overall BSC supply risk in dollars; the standard deviation of revenue at risk; and the average supplier RPI by quarter. All three indicators reveal the company is trending in the right direction.

Boston Scientific understands that being a supply chain risk management leader requires the development of the right kinds of tools and techniques to support their journey.

NAVIGATING THREATS AT BOEING

Global companies know that at any moment unpredictable events can happen anywhere in the world. These events become worse when a company is unable to determine its risk exposure. And probably no company appreciates this more than Boeing, a company with 170,000 employees in 50 U.S. states and 70 countries, and with thousands of suppliers, partners, and customers located in 150 countries. With people and operations in this many locations being affected by tornadoes, hurricanes, earthquakes, pandemics, civil unrest or terrorism, or catastrophic product failures that are major news events is not only a possibility, it is a certainty. The question becomes how severe a risk event will be and its effect on the company and its stakeholders.

Boeing created a system called ThreatNavigator to monitor diverse locations and potential risks. This system, the brainchild of several Boeing emergency management professionals, allows managers to quickly comprehend a complex situation and monitor it in real time. These personnel also envisioned a system that could track and contact Boeing employees, including those who are traveling.

An in-house team created ThreatNavigator using web technology. This tool combines internal and external information and displays it visually in a Google Maps format. External data feeds come from sources such as NC4 (a commercial information service described in Chapter 12) as well as the National Weather Service. Icons show the type of incident and use color codes to indicate the elapsed time since an incident occurred. Alerts are also sent to system users via e-mail so they can be kept up-to-date on a situation. Before the development of ThreatNavigator, alerts and information from many sources were sent to emergency responders, something that took hours to accumulate and analyze. And it doesn't take a Boeing rocket scientist to figure out that hours in an emergency represent an eternity.

ThreatNavigator, which came online in 2012, has already been used numerous times. It was first used to monitor a NATO summit hosted in Chicago to follow the actions of protestors who vowed to shut Boeing down due to the company's military support of NATO. It was also used to monitor areas affected by Hurricane Sandy during October 2012. The system also helped determine if evacuations were necessary during Colorado wildfires as well as during civil unrest in Cairo. And risk managers used ThreatNavigator to monitor the Oklahoma City area after a massive tornado hit the area as well as the aftermath of bombings at the 2013 Boston Marathon. This system also tracks medical emergencies or events at company sites daily, something that shows the system's versatility.

Other systems support ThreatNavigator. These include the DENS (Desktop Emergency Notification System), which delivers computer alerts to employees about emergencies; the DAN (Dialogic Automated Notification) system, which sends messages globally through an automated phone notification system; the BEACON (Boeing Employee Accountability Network) system, which accounts for the location and well-being of employees if a site evacuation is required; the TRIS (Travel Risk Intelligence Service) system, which monitors Boeing employees when they travel and what health and safety threats might be nearby; and other systems including an emergency 800 number and an emergency website.

These systems help keep the company and its vast network of employees safe in an unsafe world. They allow critical business operations to be maintained wherever possible during a crisis or event. By being able to react quickly to events, Boeing is at the forefront of risk mitigation, creating a confidence within the company that it will be able to deal with risk events better than ever before.

SUPPLIER RISK ASSESSMENT AT IBM

It should come as no surprise that a company known for developing innovative products and solutions for its customers would develop an innovative approach for managing supply chain risk. To grasp the importance of the supply chain to IBM, consider that the company has more than 1,800 first-tier suppliers, contractors, and manufacturing sites and more than 25,000 professionals working in its Integrated Supply Chain group. The hardware group alone at IBM buys $12 billion of production materials per year from suppliers. And the company is increasingly relying on suppliers in India, China, other Southeast Asian countries, Eastern Europe, and South American countries. As its electronics supply chain becomes increasingly complex, so too have the risks the company faces. Needless to say, the IBM of today faces more risks than the IBM of just a few years ago.

Several years ago IBM began a review of its approach to supplier risk assessment. An initiative to improve IBM's risk assessment ability started as part of an overall corporate strategy built around enterprise risk management (ERM). As part of that effort, IBM looked more closely at its risk management approaches in its supply management group. And not surprisingly, given the global nature of IBM's business, the company found opportunities for improvement.

An assessment of IBM's current state revealed that while some risks guidelines and processes were in place, risk assessment was largely a manual, and at times overly subjective, process. What constituted high risk to one individual or team might really be construed as medium risk to another. Furthermore, the approach to collecting data to support risk assessments was not especially rigorous.

The company searched the marketplace to identify commercially available solutions to evaluate suppliers in terms of financial risk, operational excellence, and product integrity. After six months the IBM team concluded that no comprehensive tool was available to satisfy its requirements. While many tools considered some aspect of what IBM was looking for, none brought everything together in a comprehensive way.

What do you do after searching the market and finding that no commercially viable approach is available to fit your needs? IBM determined it would be better off developing a tool internally supported by existing IBM products, including its Cognos analytics tool and ILOG event management capabilities. The resulting Total Risk Assessment (TRA) software tool provides automated alerts to commodity managers, purchasers, and others at IBM about potential risks in the supply chain. It also includes complex algorithms to quantitatively evaluate risk.

IBM's Risk Management Tool

IBM's primary goal when developing the TRA software was to have the ability to assess risk systematically to become more predictive and less reactive. This tool incorporates data from an existing IBM database that tracks supplier financial data and status, uses a third-party data source that provides information on worldwide news and events, and receives inputs from IBM procurement managers and others as they provide responses to a set of structured questions. It is also driven by periodic assessments of all suppliers and leverages many of IBM's in-house predictive analytic engines to calculate probabilities of risk occurrences, determine the likelihoods of events, and develop alert dashboards.

The risk management tool provides a comprehensive risk assessment and ongoing mitigation approach to protect against loss of revenue and profits by minimizing the likelihood and severity of supply chain disruptions. The tool manages and updates 13 categories of risk across five elements: country, hub, supplier, supplier site, and commodity. All new suppliers are assessed through this tool prior to awarding business with higher-risk suppliers encouraged to develop risk mitigation plans before receiving material approval.

The proprietary algorithms used to calculate risk scores from the inputs coming into the system are critical to this tool's success. The TRA tool provides a numeric score and graphically represents different risk categories as being low, medium, or high risk, thereby providing guidance to supply managers who must develop contingency or risk management strategies appropriate to those ratings. The tool provides a wide range of functionality:

- Catalogs full supplier risk exposure across multiple commodities
- Performs probability-based risk assessments
- Develops roadmap guides for risk mitigation strategies for executive approval
- Establishes control limits for each risk element
- Highlights new business processes and risk escalation pathways
- Monitors global risks against specified "risk appetite" corporate thresholds
- Updates and provides feedback to executives as risk exposures elevate
- Supports the Supply Commodity Councils as crisis situations occur

Constructing the databases to support this tool had an early benefit. When an earthquake and subsequent tsunami struck Japan, IBM was able to identify within hours the supplier sites that might be at risk from among hundreds of its Asian suppliers. Other companies required days and even weeks to arrive at a complete picture of their risk exposure. Linking the TRA system to another system that tracks on-hand inventory allowed IBM to secure supply from affected areas faster than competitors, providing IBM with uninterrupted supply to customers.

IBM has noted several initial benefits from its homegrown solution. Some of these benefits include the tool's ability to uncover multiple risks, assess the likelihood and impact of those risks, and address those risks with formal mitigation plans. The tools also provides a consistent risk management approach across all brands and commodities and provides the executive team with trends and patterns that have been revealed through systematic risk analysis.

A decade from now there will be an impressive number of risk management tools commercially available. Today, however, is a different story. Progressive companies like IBM realize they can't wait for the day when commercially available tools arrive. The time to act is now.

USING SUPPLY CHAIN MAPPING TO MANAGE RISK AT CISCO

An emerging risk management technique focuses on mapping the nodes within a company's supply chains (refer to Chapter 12 for a discussion of supply chain mapping). These nodes include production sites, warehouses and distribution centers, contract manufacturers, suppliers, and customers. Frankly, there is no shortage of nodes and entities across a complex supply chain. One leading-edge approach connects these nodes within a map with lines that indicate product volume and total price or margin, and then superimposes these with a risk index on every connection. Various tools support the calculation of the risk index, including failure modes and effects analysis (FMEA), Time-to-Recovery (T-t-R) metrics, and a Resiliency Index. Some approaches use red, yellow, or green symbols to indicate high, medium, or low risk. The resulting map, when constructed properly, produces a compelling picture.

The next step in this process involves a detailed look at the high-risk connections with the development of risk mitigation plans. However, leading companies don't stop there. Cisco, a leading producer of networking equipment, has taken its mapping process to an entirely new level. This next level integrates worldwide threats (weather, political, hazards, etc.) and superimposes any threats continuously on the map using data from companies such as NC4. Many risk management leaders, including Cisco, Bayer Crop Science, and Flextronics utilize supply chain maps and these 24/7 threat events to drive their tactical and operational risk discussions.

Here is an example of how Cisco utilizes this approach. In May 2008, Chengdu, China, experienced a magnitude 7.9 earthquake. Cisco conducted a full impact analysis that evaluated supplier sites, parts, and products in the Cisco supply chain. Within a day of the event, assessments revealed that Cisco had 20 suppliers in the affected area. Within two days of the quake, Cisco's SCRM Group initiated a crisis survey forwarded to the supplier's emergency contacts.

As a result of these efforts, two suppliers were identified as at-risk. The first supplier, a single-source supplier, represented a significant revenue risk to Cisco. This supplier was already flagged for review due to the risks associated with using a single source. In fact, Cisco had already qualified a second source. The second supplier was smaller in revenue risk but experienced significant damage to its facility. Cisco sent its Crisis Management

Team to help the supplier recover from the physical damage. The result was a faster Time-to-Recovery (T-t-R) compared with competitors. When each day means millions of dollars in lost revenues, taking risk management to a higher level means being faster and better than competitors. And at Cisco, that means taking an approach such as supply chain mapping and extending its functionality to support real-time threat visibility and risk response.

SURVIVING A NEAR-DEATH EXPERIENCE AT DELPHI

A company that can see its own demise coming tends to view the need for change a bit differently than a company that is looking at a secure future. Even before the 2008 financial meltdown, Delphi, a major first-tier supplier to the automotive industry, knew it was in trouble as it filed for bankruptcy protection in 2005. This company became familiar with just about every category of risk.

Delphi is a different company today after roaring back from bankruptcy. At a corporate level, Delphi's return on invested capital (ROIC) is now 34% compared with an 18% industry average. Profits are stronger and are now the expectation rather than the exception. As Delphi's CEO explains, "There is no commodity here. We physically monitor every piece of business we book to make sure it is equal to or better that what we have today. If the order doesn't raise the return on invested capital, then we probably aren't going to do it."[4] The company now enjoys the luxury of evaluating the effect of new orders on its ROIC measure to determine whether an order is worth accepting. You know things are better when you can pick and choose what business you want to pursue.

As part of its restructuring effort, Delphi reduced its product lines from 119 to 33, reduced its technical centers from 33 to 15 worldwide, and cut its workforce from 185,200 to 118,000. And the company now focuses on higher-margin, innovative products, particularly in a product segment it calls "active safety." It has also reduced its dependence on General Motors, which made up 54% of Delphi's sales in 2004 compared with 23% currently.

As part of its broad approach to reinventing itself, Delphi has become a leader in total cost modeling.[5] (Recall from Chapter 13 the importance of total cost analysis when managing risk.) The need to accurately understand the total cost of doing business with suppliers that are located all over the world is probably on every manager's wish list. Delphi's Cost Management group (a function within Supply Management) took the initiative to develop a desktop tool that is user friendly, requires few manual inputs, and has reduced the time required to estimate the total cost of buying a part from five days to several minutes. And perhaps best of all, the training required to use the tool takes only 40 minutes. This is important because there is a clear relationship between the complexity of a system and internal acceptance of that system.

The Cost Management group worked with logistics, manufacturing, engineering, and R&D when developing the model. Development required collecting information about transportation and logistics costs, capital costs, and currency and risk issues. An important part of any total cost model is the identification of the relevant costs that will populate the model. Perhaps most importantly, the development team took almost 18 months to validate the model's accuracy, primarily by subjecting it to real-life sourcing scenarios. (Recall from Chapter 13 our discussion of the importance of measurement validity.) The tool is accepted internally because it replaced a much more cumbersome system and has demonstrated itself to be accurate and reliable.

The risk of a near-death experience resulted in dramatic changes at Delphi. These changes clearly affected how the company looks at and responds to risk. Like many companies, Delphi realizes that a proactive approach to risk management means not having to wait for someone else to develop the tools and techniques required to be an industry leader.

MANAGING STRATEGIC RISK THROUGH COLLABORATIVE COST MANAGEMENT

Few would argue that the future of defense contractors is a bit bleak. With budget constraints affecting every Western country, this segment of the economy clearly faces strategic risks. This case highlights one defense contractor's efforts at developing a collaborative approach to managing costs.[6] The cost management process featured here moves beyond anything previously developed at this company and combines elements of value analysis, value stream mapping, project management, total cost management, Six Sigma, innovation management, early supplier involvement, and risk management into a coherent cost management process.

Over the years this company has developed various approaches, some of which are quite sophisticated, to analyze its products. While providing a solid foundation upon which to carry out detailed analyses, these approaches did not necessarily provide the insight required for altering product cost structures and winning new orders. This case describes the development of a collaborative cost management approach that was applied to a complex product that is no longer needed by the U.S. military, the company's primary customer. This pending loss of sales clearly represented a strategic risk to the company. A new, more competitive approach was needed to compete for sales from international customers.

This company decided to target an international customer to replace the expected loss of sales from its primary customer. The defense contractor believed this opportunity presented an ideal opportunity to develop a new and collaborative approach to cost management. This opportunity became the pilot program for a new process that identifies and then manages every cost element and driver within a complex product. International customers do not have pockets that are as deep as the Pentagon, making cost reduction an absolute necessity when competing for foreign contracts.

A Collaborative Approach to Cost Management

An internal cross-functional team undertook the task of documenting the current state of the product, including an extensive analysis of every cost element and driver. The company needed a complete cost picture so it could identify where opportunities to reduce costs existed. The result of this initiative was the most detailed "as is" analysis ever performed by this company. The primary objective of this exercise was to identify where design flexibility, and therefore potential cost reduction opportunities, might exist. The team examined areas as basic as quality, delivery, and operations to identify cost reduction opportunities. It also analyzed every cost component all the way to its manufacturing line cost, including machine times, labor rate and times, material costs, and costs of goods sold. Extensive cost models were developed, much more so than what could have been developed with existing methodologies. The analysis revealed that materials made up 50% of total product costs, a finding that made it clear that suppliers were going to be an integral part of this process.

After completing its current state analysis, the company conducted a two-day workshop with company engineers and designers. The first part of the workshop featured the creative generation of cost reduction ideas while the second part involved identifying potential savings. The participants also identified the cost of implementing an idea, including the cost to document an idea and verify its feasibility, as well as the cost to put an idea in place. While some ideas related to internal control and manufacturing, supplier-provided materials, as mentioned, comprised the majority of costs. This should come as no surprise as companies increasingly outsource greater amounts of value-add to suppliers, particularly in complex products like the one featured here.

At this point a decision was made to involve suppliers to generate additional cost ideas. Twenty current suppliers participated in a workshop that lasted one-and-a-half days. The company also invited potential suppliers to broaden the domain of innovation. Participants were divided into six smaller groups according to specific tracks or topics. During these workshops suppliers identified more than 150 cost-reduction ideas.

After the workshop the defense contractor evaluated the feasibility of each idea and verified whether suppliers could follow through on what they said they could do. The accepted ideas from suppliers were expected to result in more than 20% lower product costs. Moving forward, the cost management team met every week to update the supply chain's progress on these ideas.

A risk when using revised cost figures is actually achieving those figures during production. To mitigate this risk the defense contractor's spreadsheets included a risk factor column that adjusted the savings expected for items due to any uncertainty. This adjustment percentage was agreed to by a team that was familiar with the relative magnitude of potential risks across the various ideas.

Even after all this effort to lower product costs, the defense contractor did not win the new foreign contract. So, was all this worth it? The answer is a resounding yes. Perhaps most importantly, this company became familiar with a collaborative process that will help it better understand and manage costs across its current and future programs. And this company's primary customer enjoyed cost benefits through lower pricing for its remaining orders. This experience revealed in no uncertain terms the important role that suppliers play when managing supply chain costs. Looking ahead, collaborative cost management will allow this company to become increasingly competitive as it applies its newfound cost management prowess to other products and opportunities.

LEARNING ABOUT RISK THE HARD WAY AT J. C. PENNEY

Few would argue that hiring a new CEO is not a strategic decision. Dissatisfied with J. C. Penney's lackluster performance, the company's board of directors took a bold step and hired Ron Johnson, the chief executive who reinvented retailing at Apple. All the new CEO had to do was arrive at Penney's corporate headquarters in Texas on the corporate jet (which he reportedly did weekly from his home in California as he lived during the week in a high-end hotel in Dallas), spread some Apple pixie dust, sit back, and watch good things happen. What could possibly go wrong? Apparently, a lot could go wrong. After only 17 months, J. C. Penney's board ousted the CEO. This case is featured here because of its abundant risk-related lessons.

If anyone ever doubts that pricing is a strategic variable, look no further than what happened at J. C. Penney. Shortly after arriving, Johnson decided that the company's reliance on coupons and deep price discounts were simply not right for the retailer. Apparently, he also was not too fond of fixed checkout stations and cash registers. He allowed employees to wear whatever they wanted, similar to the approach at Apple where employees walk around with mobile checkout devices. Unfortunately, customers could not always figure out who was an employee or where to pay for their purchases.[7] Customer confusion soon reigned.

With minimal testing Johnson moved quickly to change Penney's business model, an act of hubris that the company may never fully recover from. He pursued an "everyday low prices" model with prices that were not necessarily the *lowest*. And, at least to Johnson, it was obvious that Penney's customers wanted new high-end brands. Bring on the new brands!

While Johnson eventually scrapped his new pricing approach, the damage was already done as customers headed for the exits. Unfortunately, new customers did not arrive to replace those who left. Repositioning a lower-end department store as one with high-end styles requires careful planning, positioning, and execution, something that did not take place as Johnson rushed to change almost everything quickly.[8] And a total misreading of the customer is usually not a good thing. As one marketing professor noted, "Ron Johnson was clueless about what makes shopping fun for women. It's the thrill of the hunt, not the buying. Women love to shop and deals are what make the game worth playing. It took billions of dollars of lost sales, lost market cap, and over a year of embarrassing performance for Johnson to realize this truth."[9] The company has since announced the closing of dozens of stores. Its very survival is even in question. The unfortunate reality is that strategic risk is the ultimate risk.

What lessons should we take from the J. C. Penney saga? First, pilot testing is a legitimate risk management approach when changing something as strategic as a company's business model. And while testing takes time, it is usually time well spent. The author of a *Harvard Business Review* article on data analytics posed an interesting question. In his article he stated, "Imagine if Ron Johnson's tenure at J. C. Penney had involved small-scale, data-driven experiments rather than wholesale changes."[10] Second, truly understanding the customer and what motivates her is invaluable. This may be Johnson's biggest mistake throughout this ordeal. Third, just because an idea worked in one industry does not make it an automatic winner in another. Apple and J. C. Penney have very different retail outlets, products, and customers.

It is also a good thing to learn from the experience of others. When Macy's acquired May department stores in 2006, a chain that relied heavily on coupons to attract customers, it decided to wean May's customers off those dreaded coupons. A year later Macy's abandoned that strategy, acknowledging publicly that pulling back on coupons was the company's biggest mistake in the acquisition. Another lesson is that when recruiting leaders, it is a good idea to make sure they believe in the organization and what it stands for. Some critics concluded there wasn't anything about Penney's that Ron Johnson actually liked. Finally, be careful that changes don't confuse the customer. J. C. Penney told customers to expect low prices, just not the lowest. Were customers really getting a deal? They weren't sure, and that did nothing to help Ron Johnson's cause. Unfortunately, some lessons are learned the hard way.

CONCLUDING THOUGHTS

A major take-away from this chapter should be the recognition that many different and creative ways are available for managing supply chain risk. Just as there is an abundance of supply chain risks, so too there is an abundance of approaches for addressing these risks. No "cookie cutter" approach is available that will be everything to everyone. The domain of risk management tools, techniques, and approaches is broad.

Excluding the J. C. Penney example, certain commonalities characterize the companies featured here. First and foremost, these companies could not wait for others to develop solutions that satisfy their specific needs. While we expect an abundance of risk management tools and systems to become commercially available over the next 5 to 10 years, and of course there are tools available now, the companies featured here feel, at least for now, they are best served by their own internal development capabilities. Second, these companies know they have not completed their risk management journey. In fact, these companies would likely admit they have merely taken a series of steps in what will be a continuous journey. Few expect supply chain risk to magically disappear anytime soon.

Something else these companies have in common is they are developing a corporate culture that understands the importance of supply chain risk management. They understand that supply chain risk management represents that place within our thought process where supply chain management and risk management intersect. And this intersection is becoming an embedded part of how each company operates. Finally, a detailed analysis at each company would surely reveal a risk champion or group that is not at all satisfied with the status quo regarding risk management. They understand that supply chain risk management is becoming a critical business process that affects a company's strategic success. They will stay at the forefront of risk management leadership.

ENDNOTES

1. Accessed from http://www.bostonscientific.com/templatedata/imports/HTML/product-safety-information.html.
2. Carson, Christine. "Navigating Threats." *Boeing Frontiers,* (September 2012): 32.
3. Carbone, James. "IBM Identifies and Eliminates Supply Chain Risk." Accessed from http://www.digikey.com/supply-chain-hq/us/en/articles/supply-chain/ibm-identifies-and-eliminates-supply-chain-risk/1507; and "IBM Details Its Total Risk Assessment Tool for Supply Management at CSCMP." Accessed from *SCDigest's On-Target e-Magazine,* October 12, 2011, http://www.scdigest.com/ontarget/11-10-012-3_IBM_Supply_Chain_Risk.php?cid=5054.
4. Bennett, Jeff. "Delphi Roars Back from the Brink." *The Wall Street Journal,* November 11, 2013: B1.
5. Siegfried, Mary. "Precision Tool Tackles Complex Task." *Inside Supply Management,* (April 2011): 24.

6. Monzka, R.M, Phillip L. Carter, William J. Markham, Robert J. Trent, Janet Hartley, Casey P. McDowell, and Gary Ragatz. "Implementing Value Chain Risk Management—Case Study Findings." *Center for Advanced Purchasing Studies*, Tempe, AZ (2012): 26–28. This company wishes to remain anonymous.
7. Townsend, Matt. "In Street Clothes, J.C. Penney's Sales Staff Goes Missing." *Business Week*, June 20, 2013, Accessed from http://www.businessweek.com/articles/2013-06-20/in-street-clothes-j-dot-c-dot-penneys-sales-staff-goes-missing.
8. Denning, Steve. "J.C. Penney: Was Ron Johnson's Strategy Wrong?" *Forbes*, April 9, 2013, Accessed from http://www.forbes.com/sites/stevedenning/2013/04/09/j-c-penney-was-ron-johnsons-strategy-wrong/.
9. Denning, Steve "J.C. Penney: Was Ron Johnson's Strategy Wrong?" *Forbes*, April 9, 2013, Accessed from http://www.forbes.com/sites/stevedenning/2013/04/09/j-c-penney-was-ron-johnsons-strategy-wrong/.
10. Davenport, Thomas H. "Analytics 3.0." *Harvard Business Review*, (December 2013): 70.

15

Future Directions in Supply Chain Risk Management

The future of supply chain risk management (SCRM) promises to be exciting, challenging, and surely a bit frightening. Something that we should have concluded by now is that the world will not suddenly become a kinder and gentler place. If we believe that to be true, and most observers are willing to bet that the world is going to become more rather than less susceptible to risk, then supply chain leaders cannot afford to lower their guard. The future will belong to those companies that not only embrace risk but also understand how to anticipate and prepare for risk.

This last chapter of the book provides a forward-looking view of supply chain risk. The first section presents a set of predictions that we believe have a high probability of occurring. The second section depicts a risk management maturity model that will help position our thinking regarding the evolution of SCRM. The chapter concludes with some managerial guidance about moving forward in terms of risk management capabilities.

SUPPLY CHAIN RISK MANAGEMENT PREDICTIONS

Let's dust off our crystal ball and take a peek into the future. While we could make dozens of risk predictions, the following are based on extensive experience and research and have a strong, although not guaranteed, likelihood of becoming a reality.

Prediction 1: Company attitudes toward risk will start to shift.
A common view among supply chain managers is that risk is something most closely associated with loss. As a result, it is something to avoid. As

companies become more confident in the economy (and that remains a bit of a wild card) as well as in their risk management capabilities, we expect risk attitudes to shift. The phrase we like to use is that companies will show a greater willingness to engage in "thoughtful" risk taking. Most of us realize that uncertainty and risk are part of a new normal. Without question many companies will become more comfortable operating with risk as they learn to compete in this new normal. Companies should become less anxious about risk, although not to the point of complacency.

Something that affects executive perception toward risk is a belief that a strong commitment to innovation, particularly radical or "blue sky" innovation is incompatible with a culture of risk management. Some will perceive that innovation leads to excessive risk, a belief that can affect new product development, strategic planning, capital allocation, and supply chain management decisions. A study by Accenture, however, found that the beta of Forbes's most innovative companies averaged just a fraction more than less-innovative peers. (A beta value is the measure of a company's share price volatility relative to the market's overall volatility.) Statistically, no meaningful relationship exists in the Accenture study between innovation and risk, at least in terms of stock volatility.

Prediction 2: Risk management will become an embedded part of supply chain management.
To date, SCRM initiatives have largely been handled separately from the normal job responsibilities of supply chain managers. As companies become better qualified in their understanding and planning for risk, we expect that risk management initiatives will become a more routine part of the supply chain management process, much as supplier audits, supplier development, supplier relationship management, and supply measurement have become a routine part of supply management.

Embedding risk management directly into chain supply management does not mean that the importance of risk management will somehow diminish. In fact, the opposite is likely to be the case. As risk management becomes a recognized part of an organization's operating culture, risk issues will increasingly be considered early on when making supply chain decisions. Supplier selection teams, for example, will not only consider a supplier's operating capabilities, but they will also routinely assess a potential supplier's financial condition as well as its risk plans and capabilities. Selection teams will also consider geographic location to ensure that a supplier is not too clustered with other suppliers or located near known hazards. The bottom line is that supply chain managers will become risk managers.

Prediction 3: Companies will emphasize risk management efforts past tier-one suppliers.

A common risk management model is to take it as an article of faith that first-tier suppliers will monitor their suppliers (which are your second-tier suppliers), and for second-tier suppliers to monitor the next tier of suppliers. Assuming this will happen regularly is a fool's proposition, to say the least.

A typical refrain when speaking with supply chain executives is that most risk management initiatives stress, and usually stop at, first-tier suppliers. A CAPS Research survey revealed that while 75% of participants report they have good risk management visibility to critical tier-one suppliers, just over 30% report having visibility to critical tier-two suppliers.[1] The need to focus on what is happening at the subtier levels of supply chains has never been greater. As one original equipment manufacturer (OEM) aerospace executive commented, "There is always some small shop out there in the sub-tiers that we just don't know about that will disrupt us." Without question the need to pursue risk management initiatives at the subtier supplier level has become a necessity.

The question becomes how to manage sub-tier suppliers from a risk perspective.

One way is to evaluate a potential supplier's risk management capabilities during supplier selection visits. To date, few companies truly evaluate how well a potential supplier manages risks in its part of a supply chain, although these companies are not shy about evaluating suppliers across many other performance dimensions. While you are assessing a supplier, why not assess how well that tier-one supplier manages its suppliers (which are your tier-two suppliers)? Another approach that will gain in popularity is mapping supply chains past the tier-one level. Graphical presentations of supply chains can provide a wealth of insight. There is simply no excuse today for not knowing what life looks like past your tier-one suppliers.

The aerospace industry provides a good case study regarding why sub-tier risk management will increasingly be on the radar screen. Figure 15.1 depicts the aerospace manufacturing supply chain structure. While each tier presents its own unique challenges, the tier-four level has started to cause concerns among industry participants. The industry has witnessed some major supplier consolidation at the tier-four level, which involves metal fabricators and casting suppliers. This consolidation is a growing

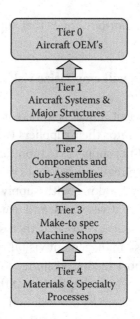

FIGURE 15.1
The aerospace supply chain structure adapted from ICF International and *Aviation Week and Space Technology.*

source of concern for customers as it creates stronger pricing power and market leverage for new, larger suppliers. One expert notes that in many cases only one or two qualified suppliers now exist for an item.[2] This is particularly troublesome as Boeing and Airbus face not only new competitors but each also has a record aircraft order backlog.

Another example of the need to understand the supply chain past tier-one suppliers involves Aston Martin, a company known for supplying the cars that James Bond drives. The company was forced to respond to complaints of the throttle pedal arm breaking. A root cause analysis revealed that a tier-three supplier in China used counterfeit material that made its way into the pedal.[3] Reports indicate that MI5 was not amused. Aston Martin announced it is planning to re-source the manufacture of pedal arms from China to the United Kingdom as soon as possible.

Prediction 4: Supply chain risk metrics will stress real-time predictive indicators.

Far too often risk management relies on historical data or a "batch" approach with periodic updating of data and information. This creates data lags that, at times, can be problematic. A performance fix may be implemented with a supplier relatively quickly, for example, but because a measurement system refreshes data only monthly, it may still show a supplier as being high risk, at least until the next update.

Many companies are searching for a better "crystal ball" for predicting risk. Rather than knowing what has happened, tools featuring predictive analytic capabilities that anticipate risk events before they occur or before an event results in more a serious loss will be highly sought after. We expect to see the increased development of supply chain operational indicators that are applied within algorithms that have predictive capabilities.

This shift toward real-time indicators will diminish the value of supplier scorecards, which are almost always backward looking. We could commit a fair amount of space here to the reasons why supplier scorecards will likely not be at the forefront of future risk management efforts. Scorecards will remain in place at most companies simply because of a reluctance to eliminate a system that enjoys a level of comfort internally and with external suppliers.

Automation and real-time updates and monitoring, including those that fill in the gaps between regular risk assessments, are a direction most companies will move toward, particularly with critical suppliers and logistics providers. Furthermore, automated data collection tools will help provide a view of risk without having to conduct formal assessments. Companies will continually search for opportunities to update the timeliness of their supply chain data and information. We expect a continued move toward real-time data and data transparency.

Prediction 5: A "pockets of excellence" risk management model will move toward an enterprise-wide risk management excellence model.
An interesting phenomenon occurs as ideas such as quality management, lean, and risk management evolve from concept to maturity. During the concept phase, internal units usually cannot wait for companywide or third-party-supported solutions, so they develop their own tools and approaches, resulting in varying levels of competency across a company. We often see internal sites or units displaying a capability that is not shared across the enterprise. The challenge becomes one of extending these "pockets of excellence" across an entire enterprise and supply chain.

Eventually, as a concept gains widespread attention, these pockets of excellence become more widespread before a company can expect to capture a sustainable advantage. This cross-enterprise standardization will be a clear indicator that risk management is maturing and growing as a discipline. In this sense, the evolution of risk management processes and capabilities will be no different from any other important process or approach that we have seen over time.

The "pockets" model also extends to the kinds of risks that a company must manage. A supply chain organization may be skilled at assessing the financial integrity of its current and potential suppliers (financial risks). However, this same organization may be less effective at detecting day-to-day issues that affect supply chain performance (operational risks). The scope of a risk management approach and the kinds of risks that a company can manage effectively will expand across an enterprise and supply chain.

Prediction 6: Total cost of ownership (TCO) modeling will become a routine part of the risk management process.

Several factors will ensure that total cost models will increasingly become a more routine part of the risk management process. First, the need to replace unit cost data with total cost data is becoming well accepted. Making important supply chain decisions with limited data simply presents too much risk. Second, a global strategy based on sourcing in emerging markets demands better data upon which to base decisions. What is the true cost of sourcing in China as exchange rates shift and the Chinese government provides incentives for suppliers to move inland in search of available labor and lower wages? What is the cost of additional complexity as supply chains become longer?

Similar to forecasting models (and some TCO models should be looked at as cost-forecasting models), risk managers will increasingly validate their data, information, and assumptions as actual inputs become available. Depending on the model, total cost models are often populated with data that are future estimates (such as estimates of a supplier's expected quality performance) or averages (such as the average nonconformance cost of a late supplier delivery). We fully expect a movement away from making sourcing decisions based on price and toward decisions based on total cost. To not do so exposes a company to excessive risk.

Prediction 7: Home-grown risk management tools will increasingly be replaced by third-party solutions.

Feel free to take this prediction to the bank. At most companies risk management tools and approaches are still in their infancy. As with many evolving concepts, early risk assessment tools are developed internally simply because external tools and applications are not yet available. Or the tools that are available don't quite fit our requirements. It is not unusual for companies to have a need, scan the marketplace, and conclude that the marketplace is lacking in its offerings. We expect third-party risk management tools and applications to proliferate as providers, particularly software providers, develop sophisticated risk management tools that respond to marketplace needs.

Thirty years ago one of the authors of this book helped to internally develop a warehouse management system (WMS) to help manage distribution centers. Why was the system developed internally? It was developed internally because no third-party tools were available that even remotely satisfied this company's requirements. How the times have changed. *Inbound Logistics* magazine regularly presents its listing of the top 100 WMS systems providers, a list that is narrowed from a pool of 300 providers!

Most companies, even those with well-recognized supply chain management capabilities, are still progressing along a risk management maturity curve. As they become more sophisticated, so too will the tools they require to manage risk. As demand increases for these tools, the marketplace will respond with an abundance of applications that will be much more powerful than what is available today. This is inevitable.

Prediction 8: Risk categories and specific risks will evolve and change, and these categories and risks will differ from industry to industry and company to company.

The risks that companies face are dynamic rather than static, particularly since many companies expect to expand geographically away from their home borders. Worldwide expansion creates greater supply chain complexity, which correlates directly with increased risk.

Evolving risks will often differ across companies and industries. A financial services firm will perceive a different set of risks to be critical five years from now, such as the integrity of its data from emerging markets, compared to companies that are heavily dependent on raw materials. Companies that rely on raw materials are increasingly concerned, for example, about the impact of rules addressing conflict minerals from Africa as well as commodity market manipulation. Other companies are concerned about evolving supplier risks regarding greenhouse gases, workplace conditions, labor issues, and the use of toxic materials.

Everyone sees risk through a slightly different prism, and that prism will continuously evolve and change.

Prediction 9: Risk management approaches will rely increasingly on anticipation and less on reaction.

Most companies recognize their proficiency, perhaps even their heroics, at responding to risk events. Unfortunately, this is not what most companies want to be known for. Most supply chain organizations would rather be recognized for their proficiency at sensing and preventing risk events rather than reacting to and mitigating risk events. Risk reaction or responsiveness, when it occurs, should be because it is the appropriate risk response rather than the default option.

Risk prevention will receive greater attention during product development and supplier selection, both of which are logical times to think about preventive activities. A word of caution is in order here. (How many times could we have used that phrase in this book?) Preventive actions taken during product design can lead to a new set of risks that must be managed. One company that is working to simplify its product designs now relies on suppliers to the point that suppliers are design partners. This arrangement, however, invites a new set of risks, particularly concerns about intellectual property (IP) ownership and becoming overly dependent on suppliers. Conversely, some suppliers are concerned about turning over their intellectual property to customers during the design process or being asked to provide customers exclusive use of new technologies or innovations.

Prediction 10: Risk management awareness will increasingly affect corporate culture, sometimes positively, sometimes negatively.

The effect that risk management has on a company's culture can be a positive or negative force as we look toward the future. On the positive side, as the language and practice of risk management becomes an embedded part of an organization's culture, personnel at all organizational levels and within all functional groups will consider risk implications when making decisions and formulating strategies. From the highest to lowest organizational levels, a continued emphasis on risk will create a healthy awareness of decision factors that may have previously been minimized or ignored. On the negative side, risk awareness can lead to risk obsession, resulting in a culture that is excessively risk averse. At that point, risk paralysis prevents a company from pursuing the kinds of activities and initiatives that support future growth.

Prediction 11: Supply chain risk will increase as companies pursue sourcing and selling opportunities in emerging markets.

For those who are dreaming about a future that features far less supply chain risk than what we see today, do not lose sight of your dream. It will likely not come true, but never lose sight of the dream. Part of the reason we think that supply chain risk will increase is that we do not expect companies to stop their incessant search for low-cost suppliers or for selling their products in emerging countries. This means that companies will be doing business in regions that are not quite as familiar or friendly as what they are accustomed to. The result of pursuing buying and selling opportunities in emerging markets is that supply chains will continue to become longer, more complex, and inevitably more risky. Within supply chain management, this will raise the need to obtain market intelligence and manage risk.

Prediction 12: Enterprise risk management and supply chain risk management will increasingly overlap.

Enterprise risk management (ERM), which most companies have practiced in some form for decades, and SCRM, which is relatively new, will increasingly be viewed as interdependent rather than independent disciplines. Part of this is due to the strategic implications associated with supply chain risks, particularly supply disruptions. And as these disruptions and supply chain events become more severe, so do their strategic implications, which is what qualifies a risk to become part of a company's strategic risk profile. As public companies fulfill their reporting requirements (refer back to the mandated reporting of corporate risk presented in Chapter 1), supply chain participants are increasingly part of this corporate reporting process. And supply chain executives are increasingly part of executive risk committees. Supply chain risk and enterprise risk will increasingly overlap.

AN EVOLVING RISK MANAGEMENT MATURITY MODEL

As we conclude our discussion of SCRM, we are convinced that as SCRM becomes a discipline, many frameworks, protocols, models, tools, techniques, and methodologies will emerge. In fact, an Internet search reveals that dozens of risk maturity models are already available across various industries and professional groups. It seems that anyone who has engaged in some aspect of risk management has created a risk maturity model. Each of these models has steps that evolve from phases such as unaware, reactive, or very basic to optimized, integrated, or advanced in terms of risk management maturity and capabilities. No shortage of models exists today.

Supply Chain Risk Maturity Model

A key roadmap and critical success factor for the SCRM journey is something we call the Supply Chain Risk Maturity Model. This new discipline is a journey, more so than a set of point-in-time solutions. Even though new solutions are emerging every day, many of which have been profiled in this book, the real key to the success will be the ability to effectively move through an SCRM Maturity Model and build expertise and knowledge that allows a company to differentiate itself from the competition. Figure 15.2 illustrates the SCRM Maturity Model. The following provides a few relevant attributes that we feel will support supply chain excellence in each stage of the risk management journey.

Visibility

Visibility and awareness of risk across the supply chain is an important first step within the risk management journey. This also relates to knowing, preferably in real time, the location of materials and assets, such as

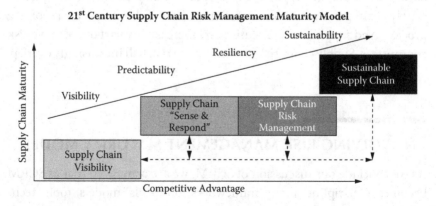

FIGURE 15.2
Twenty-first century SCRM maturity model.

transportation vehicles. During this phase most companies begin their risk assessment using emerging tools such as third-party supplier assessment tools and web-based heat maps that evaluate risk across the entire supply chain. Readers of this book have free access to a web-based Supply Chain Risk Heat Map Assessment tool that is featured in the appendix. Companies are also beginning to develop Business Continuity Plans (BCPs) for every node in their supply chains.

Supply chain leaders are starting to map their supply chains in an effort to better prepare for and respond to both demand and supply disruptions faster than their competitors. Let's not forget that SCRM is a relative game. The ability to be alerted before your competition regarding a risk event and being able to respond faster will become a critical success factor defining global excellence.

Predictability

As companies progress in their journey, we expect to see them leverage their visibility capabilities to stress test their supply chains in terms of what-if scenario planning. These exercises, utilizing network optimization and probabilistic modeling and mapping tools, provide a view into how supply chains might react to demand and supply disruptions. This insight will help companies build risk response plans.

Within this stage, companies are also leveraging third-party tools that include fraud and corruption methodologies and alerts along with new techniques, protocols, and supply chain dashboards. One result of this will be the development of supply chain risk war rooms. These war rooms, supported by new tools and techniques, will support the ability to be more proactive, relative to risk events, by identifying the risk through alerts, assessing the risk utilizing probabilistic tools, mitigating the risk through new risk protocols, managing the risk and, at times, even turning the risk into an opportunity.

Resiliency

Companies at this stage are shoring up their organizational infrastructures through corporate frameworks such as ERM; governance, risk, and compliance (GRC), and ISO risk standards culminating in Risk Centers of Excellence (COEs). Risk management leaders are also instituting new key risk indicators such as Time-to-Recovery, Value-at-Risk, and Resiliency

Indexes. These frameworks, protocols, metrics, and organizational structures provide a foundation for operational excellence in risk reduction and competitive opportunity management.

Sustainability

Companies in this stage are leveraging predictive analytics to develop "what-if" scenarios, maintaining a corporate-wide risk management framework, and building and maintaining a risk-adjusted culture supported by risk performance measures. All of this becomes a sustained part of the corporate culture. The leaders in this stage also continually assess their risk profile, evaluate their risk appetite regularly, and leverage their risk-adjusted knowledge base in order to update their supply chain risk insurance portfolio. They also achieve real-time risk event information from third-party providers and embed those alerts and information within their supply chain risk infrastructure and SCRM war rooms.

The SCRM journey is one of small steps. Each stage builds additional knowledge and expertise that needs to be codified, institutionalized, and reinforced. Supply chain leaders that are further along in this maturity mode understand that each stage in the journey corresponds to two to three years of continuous effort.

A CALL TO ACTION

A premise throughout this book is that future supply chain stability and success will be linked closely to a company's risk management capabilities. The following builds on this important premise by presenting a set of actions that provide managerial guidance.

Establish the Risk Leadership Team

It almost goes without saying that becoming a leading risk management organization will not happen without individuals who are capable of making risk management investments while demonstrating risk management *leadership*. Given the magnitude of change a typical company faces as it works to elevate its risk management capabilities, it should come as no surprise that supply chain managers must become risk management

FIGURE 15.3
Supply leadership and organization.

leaders. A continuous challenge facing organizations concerns how to develop a generation of leaders who understand how to make risk management a critical organizational competency.

One way to link leadership and risk management is to make risk management responsibility an important part of executive leadership steering committees. Leading companies, for example, have in place executive steering committees that are responsible for setting the strategic direction for their sourcing groups. Figure 15.3 illustrates the actual responsibilities of a steering committee at a leading company. (For our purposes we have added an additional responsibility related to risk management.) Wherever leadership teams exist in the corporate hierarchy, the opportunity presents itself to embed risk management responsibilities. And it is equally important to establish the risk communication hierarchy.

As an aside, leadership is not the exclusive domain of executive leaders. Whether an individual is a buyer-planner, commodity team leader, supply chain manager, or vice-president, each will have an important role to play in the evolution of supply chain risk management. Each must demonstrate leadership at the appropriate level.

Establish Risk Crisis Teams

Seasoned supply chain professionals know that "stuff" is going to happen. The question is not if something will happen, but when and how severe the consequences might be. In a crisis every second matters, making the establishment of risk management crisis teams something that should be on the agenda of any organization. These teams, staffed with the right professionals who have access to real-time data and information, are an indispensable part of the risk management process. Chapter 14 featured Boeing's approach for managing threats and crises through the use of dedicated teams.

Focus on the Risk-Management Enablers

Chapter 3 presented the idea of organizational enablers, four distinct areas that support more advanced supply chain and risk management initiatives. These enablers include a supportive organizational design; information technology systems that provide real-time or near-real-time visibility to data and information; risk-related measures and measurement systems that provide insight into potential risks as well as the effectiveness of risk management efforts; and the availability of capable human resources. Various dimensions of risk should be quantified and reported regularly to senior leadership.

The development and use of risk management tools, approaches, and strategies must become more commonplace, and that demands building a foundation of people, systems, measures, and organization that supports risk management excellence.

Assess the Current State of Risk Management Preparedness

Recall from our earlier discussion that as the concept of SCRM began to evolve, a mechanism to evaluate a company's maturity in supply chain excellence and design took shape in terms of a "spider diagram." This diagram, which profiles a company's supply chain maturity, has evolved into a supply chain risk awareness heat map. The heat map provides insight into a company's exposure to supply chain risk and the state of its risk management preparedness. It is an awareness technique that starts a dialogue about supply chain risk.

The heat map tool is used to establish an enterprise-wide picture of a company's supply chain and risks through questions-of-discovery answered by representatives from multiple disciplines. The assessment is critical for understanding a company's "as is" risk management state. The appendix of this book provides more detailed insight into this powerful and emerging approach to supply chain risk management.

Perform Risk Assessments and Develop Risk Management and Business Continuity Plans

Recall from Chapter 1 that a risk analysis or risk assessment is the process of qualitatively and quantitatively assessing potential risks within a supply chain. The objective here is to identify and categorize risks without focusing strictly on obvious risks. The risk assessment should be expansive in its scope.

After performing risk assessments, it is time to articulate risk management strategies and business continuity plans. Remember from Chapter 1 that a risk plan is a document that considers known risks and includes descriptions, causes, probabilities or likelihood of risk occurrence, costs, and proposed risk management responses. The risk plan should be widely communicated with regular updates and progress reports. And recall from Chapter 7 that business continuity planning is the process of planning for and implementing proactive, preventive procedures that are designed to enable continuous operations of critical business processes and functions.

Gain Visibility across the Supply Chain

As we refer back to the Supply Chain Risk Maturity Model, the importance of supply chain visibility is clear. It is hard to envision becoming a risk management leader without a strong awareness of what is occurring across the supply chain. And visibility starts our risk management progression toward the more advanced phases of SCRM. A question logically becomes where we need supply chain visibility. A partial visibility listing includes the following:

- Location and performance of subtier suppliers
- Current and predicted operational performance of key tier-one suppliers
- Updates of supplier financial health
- Real-time location of materials and transportation assets
- Employee movements worldwide
- Hazard risks such as conflicts and weather
- Changes to supply and customer markets

The importance of the information technology enabler cannot be overestimated when striving for supply chain visibility.

Benchmark Risk Management Practices against Industry Leaders

Benchmarking is the continuous process of measuring products, services, processes, and practices of a firm against world-class competitors or those companies recognized as industry or functional leaders. Why benchmark risk management practices? In short, benchmarking helps accelerate the development of best practices from any industry, stimulate and motivate those who must implement benchmarking findings, break down ingrained resistance to change, identify technological breakthroughs from other industries, and provide valuable professional contacts. Benchmarking is also a way to create a corporate culture that understands risk.

Organizations that fail to benchmark demonstrate certain undesirable characteristics. They tend to be reactive to change while pursuing internal, evolutionary change based on historical performance. They also tend to demonstrate a "not invented here" syndrome. On the other side, organizations that actively benchmark display a different and more desirable set of characteristics. These organizations tend to be forward thinking, proactive, and receptive to new ideas that accelerate the rate of change. And, they continuously search for industry best practices. As you might be able to tell, we are proponents of performance benchmarking.

Develop or Obtain the Tools, Techniques, and Risk Protocols

Algorithms using sophisticated predictive analytics and big data will increasingly become the norm rather than the exception in risk management. We expect an abundance of new tools and techniques to become available. As with any tool, the key will involve understanding how to use it properly.

CONCLUDING THOUGHTS

While this book's risk management journey is coming to an end (we are, after all, in the conclusion section of the last chapter), it is probably safe to say your risk management journey is evolving or even just beginning. We would like to close by reiterating some important observations made in the preface to this book (especially since many readers skip the preface):

- The financial impact of supply chain disruptions can be devastating but is often not understood until it is too late.
- The supply chain management profession has become too comfortable with the deterministic models and tools developed over the last 35 years.
- Supply chain risk management is an evolving discipline and will remain so for the foreseeable future.
- Supply chain strategies driven primarily by cost management and delivery improvements are no longer comprehensive enough.
- Hard return on investiment for risk management initiatives is a hard sell.
- Social media is the new risk wild card.
- The risk ledger has two sides—negative consequences and potential opportunities.
- Supply chain risk is making it to the big leagues in terms of corporate visibility and reporting.
- Risk heroics must give way to risk prevention wherever possible.
- We need to take a broader rather than narrower view of supply chain risk management.
- Global supply chain risk is increasing, not decreasing.

In the end, supply chain risk management starts with an enlightened leadership that understands how to manage this thing called supply chain risk.

ENDNOTES

1. CAPS Research Team. "Supply Risk Blind Spots." *Inside Supply Management*, (November/December 2013): 34.
2. Michaels, Kevin. "Revolution for Below." *Aviation Week and Space Technology*, September 9, 2013: 18.
3. Larson, Christina, "James Bond's Sports Car Has Chinese Supply-Chain Problems." *Business Week*, February 5, 2014, retrieved from www.businessweek.com.

Appendix: The Supply Chain Risk Assessment Tool

As we mentioned in Chapter 12, as the concept of supply chain risk management began to solidify into a body of knowledge and then courseware, within the framework of an enterprise-wide risk management approach (ERM), several major tenets developed into a set of questions-of-discovery regarding the maturity of the supply chain. As these supply chain tenets began to solidify, a mechanism to review, evaluate, and position a company's maturity in supply chain excellence and design took shape in terms of a "spider diagram" or heat map profiling a company's supply chain maturity, which then relates implicitly to the risk associated with that maturity. The heat map is a set of about 100 questions-of-discovery across 10 tenets of an end-to-end supply chain environment. A graphical illustration of the heat map appeared in Chapter 2. The answers to these questions-of-discovery provide a first-of-a-kind glimpse into a company's supply chain risk. The spider diagram, featured in Figure A.1, profiles color-coded levels of risk that correspond to high, medium, and low risk.

HOW YOU MIGHT UTILIZE THE TOOL

The heat map provides, perhaps for the first time, a glimpse into a company's risk within their entire supply chain. As shown in Figure A.1, the tenets are leadership, balance scorecards, sales and operations planning (S&OP) processes, IT systems, supply chain techniques, demand management, industry practices, manufacturing, supply base, and logistics. The tool is an awareness technique to begin the dialogue covering supply chain risk. This figure has an outer ring (a red ring when using color-coded systems), which means any answer outside the red ring equates to a very high risk. Any answer between the red ring and the yellow ring (again, in color-coded systems) equates to a moderate risk. And any result inside the yellow ring implies a lower risk. We placed an actual set of

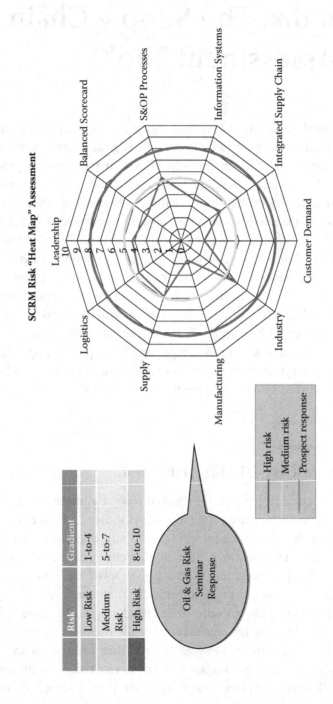

FIGURE A.1
SCRM heat map assessment.

scores from an oil and gas company headquartered in Asia. The reason the tool resonates is that many companies still have a tremendous amount of suboptimization within organizational silos and a lack of supply chain visibility.

The heat map tool is also utilized during supply chain risk assessment engagements with companies. The tool is used at the beginning of an engagement to establish an agreed-upon enterprise picture of the company's supply chain and risks through questions-of-discovery answered by leaders from various disciplines. Next, a detailed risk assessment for each node within the supply chain is exercised with the culmination of a final review of the supply chain risks using the heat map tool.

WALKING THROUGH THE QUESTIONS-OF-DISCOVERY

As mentioned, there are about 100 questions inside the tool. The questions are indicative of supply chain best-in-class practices from more than 25 years of operational executive experience and 15 years of executive supply chain consulting around the globe. Our basic premise, reinforced by many research organizations and top-tier consulting companies, is that as the supply chain becomes more mature, risk to the organization is diminished. Remember, risk is not eliminated, but rather reduced.

As you answer the questions, the tool will calculate a rating within the spider diagram at the bottom right end tab of the spreadsheet. When completed, the user will have a full-picture image of inherent risks throughout an entire supply chain. The risk spider diagram can be used to begin or continue the supply chain risk dialogue.

HOW TO ACCESS THE TOOL

As a book owner and professional interested in SCRM, you will have access to the supply chain risk assessment tool, at no cost, for 45 days after your initial download. To access the tool follow the instructions in Figure A.2.

1. **Participants**
   a. Download using this URL →
      http://www.mysnapps.com/survey/SCMRiskAssessmentWS.xlsx
   b. Password to open file: **go_journey**
   c. Enter your company name (or any other identifier) in the box labeled "enter Company Name." It will appear on each Rating tab and in the legend of the Heat Map.
   d. Enter your Valuation Ratings in the "Participant Rating" column of each Rating tab. Your results will appear automatically on the Heat Map.
   e. You can print your Heat Map and/or your ratings using the Print menu.
   f. To create a pdf with all of your ratings and your Heat Map, use the "Save as Adobe pdf" option. You must convert the entire workbook (ignore the warnings about bookmarks, links and protected sheets by clicking "Yes" or "OK" in each case).
   g. Questions should be directed to Greg Schlegel at schlegel01@earthlink.net.

FIGURE A.2
Accessing the heat map tool.

Index

Printed in the United States
by Baker & Taylor Publisher Services